T0074637

Smart Manufacturing

The manufacturing industries remain the foundation of local, regional, and global economies. Manufacturing plants operate in dynamic markets that demand upgrading with transformational technologies for maintaining profitability, competitiveness, and business sustainability. Yet most manufacturing plants currently use technologies that are no longer competitive, and industry leaders face an overwhelming array of operational challenges that require agile and enhanced transformational solutions.

This book offers manufacturers effective strategies and tools for the adoption and implementation of advanced operational technologies to ensure long-term innovation, efficiency, and profitability.

- Covers advanced automation integration in manufacturing, including digitization, AI, machine learning, IIoT, and cybersecurity
- Describes innovation, development, and integration of control technologies for sustainable manufacturing
- Explains how to upgrade existing manufacturing plants for the global market
- Shows how to apply emerging technologies including asset optimization and process integration for product lifecycle improvements, plant operation and maintenance enhancement, and supply chain integration

This book serves as a strategic guide to applying advanced operational technologies for engineers, industry professionals, and management in the manufacturing sector.

Commercializing Emerging Technologies

Series Editor Hebab A. Quazi

This series features books that focus on commercialization of new and emerging technologies. Topics of interest include automation, security, sustainability, smart manufacturing, nanotechnology, and process intensification and are aimed at researchers and industry professionals interested in advancing their work from the lab to the commercial marketplace.

Commercializing Nanotechnology
Hebab A. Quazi

Smart Manufacturing: Integrating Transformational Technologies for Competitiveness and Sustainability
Hebab A. Quazi and Scott M. Shemwell

For more information about this series, please visit:
https://www.routledge.com/Commercializing-Emerging-Technologies/book-series/CRCCET

Smart Manufacturing
Integrating Transformational Technologies
for Competitiveness and Sustainability

Hebab A. Quazi
and
Scott M. Shemwell

CRC Press
Taylor & Francis Group
Boca Raton London New York

CRC Press is an imprint of the
Taylor & Francis Group, an **informa** business

Designed cover image: © Shutterstock

First edition published 2023
by CRC Press
6000 Broken Sound Parkway NW, Suite 300, Boca Raton, FL 33487-2742

and by CRC Press
4 Park Square, Milton Park, Abingdon, Oxon, OX14 4RN

CRC Press is an imprint of Taylor & Francis Group, LLC

© 2023 Hebab A. Quazi and Scott M. Shemwell

Library of Congress Cataloging-in-Publication Data
Names: Quazi, Hebab A., author. | Shemwell, Scott M., author.
Title: Smart manufacturing : integrating transformational technologies for competitiveness and sustainability / Hebab A. Quazi and Scott M. Shemwell.
Description: First edition. | Boca Raton, FL : CRC Press, [2023] |
Series: Commercializing emerging technologies |
Includes bibliographical references and index.
Identifiers: LCCN 2022050813 (print) | LCCN 2022050814 (ebook) |
ISBN 9780367742928 (hbk) | ISBN 9780367742935 (pbk) |
ISBN 9781003156970 (ebk)
Subjects: LCSH: Production engineering. | Manufacturing processes.
Classification: LCC TS155 .Q47 2023 (print) | LCC TS155 (ebook) |
DDC 670—dc23/eng/20230130
LC record available at https://lccn.loc.gov/2022050813
LC ebook record available at https://lccn.loc.gov/2022050814

ISBN: 978-0-367-74292-8 (hbk)
ISBN: 978-0-367-74293-5 (pbk)
ISBN: 978-1-003-15697-0 (ebk)

DOI: 10.1201/9781003156970

Typeset in Times
by codeMantra

Contents

Preface

The National Institute of Standards and Technology (NIST) in the United States defines Smart Manufacturing systems as "fully–integrated, collaborative manufacturing systems that responds in real-time to meet changing demands and conditions in the factory, in the supply-chain network, and in the customer needs." Smart Manufacturing is an ecosystem that enables factories, warehouses, and supply-chains to be fully integrated and agile to fulfill customer demand for products or services. Sometimes, these activities are labeled as "Industry 4.0" or "Digital Transformation."

Amongst professional organizations, ISA (International Society of Instrumentation) has taken the lead in defining "Smart Manufacturing" for the industries. ISA is focused on creating industry standards for performance, safety, and security. These standards are for enterprise-wide control-system integration, safety, and security. ISA has taken lead in developing new standards that would be needed as Smart Manufacturing and Industrial Internet-of-Things (IIoT) technologies progress. With these objectives, ISA has now the "IIoT and Smart Manufacturing" Division with several technical committees. These committees are defining different aspects of "Smart Manufacturing" so that appropriate standards can be developed. The ASME (American Society of Mechanical Engineers) has now a "Smart Manufacturing" team looking into the technical details that would help create the required standards. The IEEE (Institute for Electrical and Electronics Engineers), ASME , and the AIChE (American Institute of Chemical Engineers) are working to support different "Smart Manufacturing" methodologies and standards for the manufacturing industry.

Most of the Smart Manufacturing initiatives are to capture the opportunities in automating manufacturing operations for improving agility, increasing efficiency, enhancing quality, and reducing barriers to market entry. In capturing these opportunities, it requires use of the emerging and transformational technologies. These activities require the industry to be better connected, more intelligent, and most importantly be dynamic systems. There, action items include minimizing risks, improving reliability, and optimizing operation. The prime focus of any Smart Manufacturing is to make the business profitable, competitive, and sustainable.

Transformational Technologies that have potentials include Digitization, Artificial Intelligence, Machine Learning, Digital Twin, Industrial Internet of Things, and Cyber Security. Appropriateness of these technologies varies in the application of Automation, Optimization, Energy Conversion and Conservation, Resource Utilization, Assets Utilization, Quality Management, and Supply-Chain Integration.

Application of these transformational technologies influences business profitability, competitiveness, and sustainability. Smart Manufacturing know-how will assist the industry appropriately. The approach to processing, mechanical operation, raw-materials conversion, products-specification, and supply-chain operation are all keys to Smart Manufacturing success. Technological innovation plays an important role in the success of Smart Manufacturing. Certain industries are in transition, and data-focused innovation sometimes brings enormous benefits to manufacturing. It is very important to have the automation professionals with appropriate knowledge to bring improved benefits from these transformational technologies. These steps can assure the manufacturing industry profitability, competitiveness, and sustainability. Appropriate digital transformational know-how can benefit the ever-changing digital landscape. Knowledge of the latest techniques in IIoT, AI, data analytics, and simulation along with edge computing and cloud computing can take the manufacturing industries to the next generation. Also, the knowledge about the dynamic world of automation including cybersecurity would be essential for protecting the plant data from external threats.

Strategic operation of Smart Manufacturing facility is essential for staying profitable, competitive, and sustainable in today's highly dynamic global economy. Supply-chain operation and logistic planning are also very critical to survive in this competitive business environment. For these reasons, the manufacturing businesses must have well-trained management and operating teams.

A well-prepared roadmap to navigate the industry for a successful and sustainable future is essential. The strategic direction of the industry must be very clear and well defined. Sustainability initiatives need to be collaborative with customer behavior, where possible, and be agile and flexible with skilled talents.

Acknowledgments

The authors of this book are indebted to industry leaders and to their key staff for sharing their desire in upgrading and modernizing their plants for competitiveness and sustainability. For this, the requisite tools and techniques are not generally available in the marketplace. The industries require clear guidelines for adopting and integrating transformational technologies for profitability, competitiveness, and sustainability. The book provides the industries step-by-step methodologies for digitizing their plants and adopting and integrating transformational technologies and techniques to make their plants productive, profitable, competitive, and for sustaining growth year after year.

We are thankful to Professor Peter Drucker who said, "Innovation is the specific instrument of entrepreneurship… the act that endows resources with a new capacity to create wealth." In the adoption and incorporation of transformational technologies in Smart Manufacturing practices, innovative approaches are necessary for profitability, competitiveness, and sustainability.

Authors

Hebab A. Quazi, PhD, has been the Chief Operating Officer of MARTECH International Inc. for over 40 years. He provides leadership to US federal, state, and city government programs and to manufacturing industries within the United States and internationally. He led programs funded by the World Bank, the United Nations, the Asian Development Bank, the US Department of Defense, and the US Department of Homeland Security. Dr. Quazi was also the Vice President of Operations for the South and East Asia and North and South America of a European technology company. Dr. Quazi earned a PhD degree from England as a Commonwealth Scholar. He also completed a 3-year graduate business education at the University of California. He is a Registered Professional Engineer in the State of California. Dr. Quazi is very active with professional organizations such as AIChE, ASME, and ISA. Currently, he is the Chairman of the ASME Energy Storage Committee and the Vice Chairman of the ISA Digital Twin and Simulation Committee. He has written several books and technical papers.

Scott M. Shemwell, DBA, Managing Director of the Rapid Response Institute, is an authority and thought leader in field operations and risk management, with over 30 years in the energy sector leading turnaround and transformation processes for global S&P 500 organizations as well as start-up and professional service firms. He has been directly involved in over $5 billion in acquisitions and divestitures as well as the management of significant global projects and business units. He is a leader in the use of data and information enabling Operational Excellence for over 20 years. Dr. Shemwell holds a bachelor of science in physics from North Georgia College, a master of business administration from Houston Baptist University, and a doctor of business administration from Nova Southeastern University.

1 Smart Manufacturing

Hebab A. Quazi

1.1 INTRODUCTION

Smart Manufacturing needs to focus on business profitability, competitiveness, and sustainability. Most often, these are the results of imagination, creativity, and innovation. It involves product and market restructuring and adoption of transformative technologies utilizing manufacturing data intelligently. Digitized operational processes are the foundation of Smart Manufacturing. It demands competency-directed and profitability-focused operation.

Innovation is behind a lot of operational changes that are needed in the manufacturing industry. The manufactured products in the future most likely are the results of innovative efforts. The pressure will continue for innovative products. Recent innovative research and development work opened the door for innovative manufacturing operations for new products including nanotechnologies. Nanotechnology brings tremendous potential for new applications requiring innovative manufacturing processes.

Emerging technology integration into existing manufacturing operation requires predetermined roadmap. Most often, these roadmaps include technology readiness assessment, prototyping, and testing, as appropriate. The roadmaps will include manufacturing-process integration and operational-readiness verification.

The benefit of digital transformation is known for quite some time now. But, the costs, schedule to implementing, and return on investment are the challenges. Some manufacturing industries have taken cautious steps and ventured into Smart Manufacturing operation. The adoption of advanced digitalization in industry is continuing.

1.1.1 PRODUCTIVITY

Business entities producing products in manufacturing plants sell in competitive markets. The products with appropriate quality are sold in a competitive market at a price paid by the customers. The satisfied customers come back and buy again, generating revenues for profits, year after year. These products are to be selling in a competitive environment. Key criteria in this competitive market are the repeat sale providing profits year after year. The profits must be on a long-term basis to provide sustainability.

For transformational technologies to survive and sustain in manufacturing, operation must work within strategic plans (market entry plan, sustainability plan, risks mitigated, intellectual property protected, regulatory compliant, and prudent business plan readied), fundraising plan, and product launching plan. Targeting traditional markets is relatively easy including Agriculture/Food, Bio-Engineering, Defense, Energy, Engineering, Environmental, Healthcare, and Transportation. Advanced manufacturing processes including adsorption, desorption, separation, and reaction are sometime challenging. Also, industry best practices are to be utilized where applicable.

DOI: 10.1201/9781003156970-1

1

These best practices can include program planning, budgeting, and controlling; quality assurance and control; scheduling and schedule control; cost estimating and control; testing and validation; and prototyping and scale-up.

1.1.2 DIGITIZATION

Currently, certain manufacturers are consolidating their relevant plant data and engaging IT/OT analytical specialists to develop their roadmap(s) to planned success. These roadmap(s) do provide appropriate direction(s) for their future business activities. If these activities are not done properly, it can lead to wasteful efforts. The manufacturers sometime engage specialty firms for bringing in external expertise and know-how.

Manufacturing technologies can provide competitive advantages and sustainability. Each industry has its own strategy in owning and protecting its technological assets and data. These manufacturing companies navigate their pathways in managing their proprietary data and/or developing further technological reach for their business success. Sometimes, intra-company exchange brings benefits for short-term and/or long-term benefits. Information sharing benefits development of cybersecurity and regulatory compliance strategies.

Many manufacturing companies are still resisting adopting digitization. Sometimes, the information on Return-on-Investment (ROI) for certain industries are not easily available. Most likely, manufacturing industries will have to develop their ROI for their digital-transformation projects and sometime may have difficulties in securing management approval to proceed. It is suggested that each company prepares their own "Strategic Plan" for transformational technology adoption assuring profitability, competitiveness, and sustainability. It requires a focused and dedicated team with a leader having adequate experience. If and when required, the manufacturing company engages an outside resource company for assistance or guidance.

Recent experience has shown that Industry 4.0 is a good roadmap for growth and profitability. An industry forecast shows that Industry 4.0 roadmap would give about 15% annual growth rate from 2020 to 2027. However, the growth rate for transformational technology adoption in Smart Manufacturing roadmap should result in much higher annual growth with more focused competitiveness and sustainability.

Securing large profitability with better competitiveness and extended sustainability would require prudent integration of several business activities. These include strategic plan developed with careful consideration of appropriate data management with adequate security. At the early stage, it is difficult to make enough small changes in digitizing operation. Most likely, manufacturing plant digitizing operation should begin with consulting an external knowledgeable specialist.

Accelerating the operational efficiency in the plant could be a vital step in digitization process. At the beginning, steps like adding new sensors in appropriate locations could be a prudent approach. Accuracy and validation of collected data can be another small step to success. Adding a historian to the plant having DCS or PLC knowledge would be a good approach. Also, considerations should be given on how to automate a particular section of the plant that requires digitization not yet automated. In the planning for digitization, the future operational plan needs to be spelled out. The plant expertise-development plan should be included in the digitization plan. Finally, the costs for digitization should be identified and ROI be calculated. The digitization team's performance needs to be appraised periodically. Also, the question of any replacements (hardware/software) with costs and timing would have to be addressed in the plan. Centralized support system for the proposed changeover is critical to digitization program. Digital transformation can and (most likely) will make significant impact on the operation of the plant. However, the digitization process will make the plant more efficient, profitable, and sustainable. To make these happen, the plant needs to make concerted efforts to make presentations to the Operation and Management Teams, Board and Supply-Chain Leadership, as appropriate.

1.1.3 IT/OT/IIoT

There are many sources of IT/OT data in a manufacturing plant. But IT/OT convergence in a plant require right strategy and tools. It is prudent to use the available tools first for the convergence. Adding additional/or new tools should be considered carefully. Keeping the existing historian saves potential problems and allows continuation of production without any problems. Manufacturing resources are at risk when changes are made. Replacing or modifying existing data systems and real assets is expensive and likely to introduce risks. Appropriate risk assessment and mitigation should be undertaken before making any changes. An appropriate and knowledgeable team of specialists is a key to successful IT/OT convergence program. Transformational technologies in the IT/OT sector require addressing digitization, artificial intelligence, machine-learning, digital-twin, industrial Internet-of-Things, and cybersecurity.

Industry managers tasked with technology commercialization need to consider prototyping expenditures most likely from outside sources and then obtain approval from the company top management. Before obtaining management approval, these cases must have evaluation of short-term and long-term benefits with risks mitigation. Situational analysis will have clear statement on current status, proposed solution, alternatives, if any, and risks analysis with recommendation for risk avoidance or mitigation. The business case needs to be described in detail so that the cost associated with each step can be estimated. The capital expenditure and their justifications should be prepared for investors or management consideration and approval. The proposed solution also should be broken down into details so that it is easy for the approval authorities to understand the measurable benefits. The proposal should identify clearly the value-added benefits, including net profits that can be achieved. Also, estimates of time for each step and their benefits analysis are necessary. Alternatives considered, if any, must be clearly stated along with their associated costs and potential benefits and risks. It is very important to demonstrate that fair and sound evaluation can be conducted by the management.

1.1.4 CYBERSECURITY

Cyberattacks are increasing in manufacturing. The attacks to manufacturing industries have increased worldwide, about threefold in 2021. The challenges to their network specialists for these industries have increased tremendously. According to the research conducted recently, the adoption of micro-segmentation concepts has earned great reputation against cybersecurity.

An overwhelming majority of the network-security professionals in the industry credit micro-segmentation techniques as the prudent approach to cybersecurity. Still, some industries lag behind in micro-segmentation. Manufacturing industries manage large quantities of operational data making them vulnerable to intellectual-property theft. Micro-segmentation of operational processes and data makes them less vulnerable to cyberattack. Currently, it is considered a prudent approach to adopt to micro-segmentation in the manufacturing operation.

1.1.5 OPERATIONAL STRATEGY

Effective application of transformational technologies for competitiveness and sustainability requires end-to-end solution. The adoption or insertion of transformational solutions in a manufacturing plant require (1) identifying and defining the problem, (2) developing the solution, (3) describing competitive advantages, (4) selecting the transformational technology for the application, (5) assessing technology's current readiness level, (6) developing the implementation plan, (7) arranging financing, (8) building a prototype for demonstration, if required, and then (9) manufacturing and testing the market. It is suggested that the proposed roadmap is used, where possible.

1.2 CHALLENGES AND OPPORTUNITIES

Transformational-technology applications for commercial products include sectors such as agriculture and food, bio-engineering, defense, electronics, energy, engineered materials, environment and safety, and medicine and healthcare. In addition, innovations have prompted us to look at a large number of applications for transformational technologies. Sometimes, steps involve innovation, adoption, and/or integration of emerging transformational technologies. These steps include technology readiness assessment, prototyping and testing, manufacturing, sustainability planning, risks assessment and mitigation, market testing, financing, and customer acceptance. Commercialization process requires end-to-end solution.

1.2.1 IDENTIFYING CHALLENGES

The challenge begins with understanding and identifying the road map the technology product will have to go through. These include (1) discovering a consumers' need or application, (2) defining the application clearly and estimating the market potentials, (3) creating the solution, (4) describing the competitive advantages of the solution offered, (5) selecting an appropriate technology or technologies, (6) assessing the current readiness level of the technology selected, (7) developing a sustainable commercialization feasibility plan, (8) arranging funds or financing, (9) building a prototype for demonstration, (10) manufacturing few sample units for market testing, and (11) testing the market and receiving customers' comments.

1.2.2 TARGETING OPPORTUNITIES

The market sectors that have high potential for future applications as commercial products are addressed here. Additional opportunities include sectors like defense, electronics, energy, engineered materials, environment and safety, and innovative manufacturing. Potential opportunities can be identified by "Viewing Broader Markets" that exist under artificial intelligence (AI), climatic-change challenge, digital transformation, innovative infrastructures, and smart cities. Adoption and/or integration of emerging technologies require technology readiness assessment, prototyping and testing, manufacturing, sustainability planning, risks assessment and mitigation, market testing, financing, and customer acceptance.

Commercialization process requires end-to-end solution. Commercialization of a product will have to go through the following steps:

1. Discovering consumers' need or application.
2. Defining the application clearly and estimating the market potentials.
3. Creating the solution.
4. Describing the competitive advantages of the solution offered.
5. Selecting an appropriate technology or technologies.
6. Assessing the current readiness level of the technology selected.
7. Developing a sustainable commercialization feasibility plan.
8. Arranging for funds or financing.
9. Building a prototype for demonstration.
10. Manufacturing few sample units for market testing.
11. Testing the market and receiving customers' comments.

The world of manufacturing is vast. One requires viewing the broader market potentials beyond the current interests that exist in the technology community. Certain applications of technologies may be overlooked initially, but these will be tagged once their sustainable benefits are recognized. One can visualize the technologies that can currently impact our everyday life. There are applications with huge potentials, but they need a small push to make them into commercially available products. Currently, some of these ideas are confined in the research and development stages but could become commercial products in the near future. Sometimes, opportunities can be created or developed.

1.3 SMART MANUFACTURING TECHNIQUES

1.3.1 TRANSFORMATIONAL TECHNOLOGIES

Adoption of transformational technologies begins with the selection of appropriate techniques in developing the strategy, plans, and programs. First, it is important to assess whether the technology is ready for the commercializing process. In this case, Technology Readiness Assessment (TRA) methodology and techniques are to be considered. Once the technology is considered ready to move from research labs to prototyping, then consideration should move to prototyping design and financial requirements. Most often, the industry needs to borrow money for this phase. Private and/or commercial lenders require that the technology developer convince that they are capable of repaying back. Major steps include selecting appropriate manufacturing processes that would increase the viability of commercializing the venture.

Once the technological steps are established, then it is very important to protect the intellectual property rights through patenting.

The TRA guidelines are systematic approaches to assist in establishing the current status of a technology, prior to proceeding with commercialization. The TRA assessment methodology is a tool that establishes the current maturity level of the technology and helps estimating the remaining work that should be carried out before the technology is considered commercial. These guidelines can be used to estimate the capital and timeline requirements for commercialization. The guidelines require that (1) the prototype is designed, built, and tested in a relevant environment; (2) demonstrated that the prototyping was done in an operational environment; (3) system was proven to work in its final form under desired conditions; and finally (4) system that has successfully performed repeatedly in a real-world environment.

1.3.2 PROTOTYPING

Scientists, engineers, and managers tasked with technology commercialization are challenged with prototyping. In addition to addressing technical issues, prototyping requires expenditure of capital that must be raised from outside sources requiring approval from the company top management. The challenges include establishing convincing value proposition for needing the prototyping steps and then developing the financial justification in a business case for acceptance by the investor or by the company top management. The business case must include short-term and long-term benefits with risks assessment and mitigation decisions. This should include at least situational analysis and financial justification.

Situational analysis should provide clear statement on current status, proposed solution, alternatives, if any, and risks analysis with recommendation for risk avoidance or mitigation. Current situation analysis is the first step in finding the solution. Current situation should be defined first in detail so that the costs associated with each step can be estimated. This is essential so that capital expenditure and its justifications can be prepared for investors or management consideration and approval. The proposed solution also should be broken down into details so that it is easy for the approval authorities to understand the measurable benefits.

The capital-expenditure proposal identifies clearly the value-added benefits (including net profits that can be achieved) and the time that it will take before the benefits can be realized. Alternatives considered, if any, must be clearly stated along with their associated costs and potential benefits and risks. It is very important to demonstrate that fair and sound evaluations were conducted. All potential solutions, particularly requiring capital-expenditures, have associated risks. The risks must be assessed clearly, and the degree of risks ranked. Risks analysis needs to have, where appropriate, recommendations for risk aversion or mitigation.

Financial analysis with justification is the core function of the prototyping business plan, and it must include ROI estimate on the capital requested for prototyping. The justification normally has two parts: (1) capital investment requirements including expenses and (2) financial benefits including ROI or pay-back-period. The best case scenario would show a large return with small investment. But, in reality, the top strategy would be to select the one with most compelling benefits with good return that requires minimum investment. Consider the measurability of the proposed ROI as the first pick. To increase the probability of success in securing the approval of the prototype funding, it is important to identify the hardware that can be reused in the operation after prototyping.

Three options for prototyping are (1) restructuring existing prototyping methods, (2) increasing existing prototyping activities, and (3) rapid prototyping. Selection of appropriate prototyping option begins with a search for previous experience in outsourced or/and in-house prototyping. Collection of historical data for any options of prototyping is necessary during the last 3 years, if available. This review provides a baseline for all potential options.

1.3.3 TRANSFORMATIONAL-TECHNOLOGY COMMERCIALIZATION

Transformational technologies are developed, tested, and validated for use in commercial manufacturing plants. Prudent evaluation of the options is required before selecting one for commercial application. Transformation technologies that have potential include informational technologies such as artificial intelligence, machine learning, and Industrial Internet of Things (IIoT).

Currently, the manufacturing processes and techniques that have high potential for consideration are combustion, deionization, electro-spray, flow-synthesis, IR spectrometry, laser deposition, morphology manipulation, nanofiltration, separation, process synthesis, pyrolysis, and vortex mixing.

1.3.4 INTELLECTUAL PROPERTY PROTECTION

Before completion of the research and development phase, the question whether the results would be patentable or not should be initiated and discussed within the organization. Potential solutions to competitiveness and sustainability require capital investment having associated risks. A program team should be assembled from the research and development, manufacturing, marketing, management, and legal departments as a minimum to provide the direction and guidance. The technology concept being new is not enough to convince the examiner from the patent office. It is very important to understand the local patent office requirements. If possible, it would be good to understand the requirements for the regional and/or global patent rights.

The patent application should not only describe the invention but also differentiate it from similar concepts already patented. The patent rights when granted will also assign the territory and the duration it is valid for. The required fees for patent application can vary from country to country, region to region, as well as for global coverage.

1.3.5 RISK MANAGEMENT

The Risk Management System (RMS) can be different for each manufacturing facility based on operational processes used in each. In the United States, manufacturing facilities are required to have Injury and Illness Prevention Plan (IIPP). Each facility is required to provide RMS and IIPP training to the employees. Usually, the manufacturing industries use outside resources for risk management and preparation of RMS and IIPP.

The industry guidelines provide the basis for consistent management of manufacturing facility operational hazards and risks to the plant personnel. Sometimes, each manufacturing company develops its own guidelines and practices. These guidelines are practiced in all of the facilities of the organization for consistency. The manufacturing industry sets their guidelines frequently after appropriate reviews of their practices. The Program Implementation Schedule is set by the industry.

The risks must be assessed clearly, and risks ranked. Risks analysis needs to have, where appropriate, recommendations for risk aversion or mitigation. The steps that are required as a minimum include (1) evaluating the market (size, trend, and sustainability), (2) identifying and evaluating competitions (competitive products, market shares, pricing trend, etc.), (3) establishing the strategy for market entry, (4) identifying appropriate market entry methodologies, (5) matching own company strengths and weaknesses, (6) identifying market (local/regional/cultural) preferences, if any, and (7) checking for special situations such as economy of scales, logistic challenges, and regulatory climate, as applicable.

1.4   COMPETITIVENESS AND SUSTAINABILITY

Commercializing technology—from Laboratory to Market involves innovation, adoption, and sometimes integration in an existing business process. According to Steve Jobs, "Innovation has nothing to do with how many R&D dollars you have... It's not about money. It's about the people you have, how you're led, and how much you get it." Dale Carnegie's guide to innovation includes finding ideas, finding solutions, finding acceptance, implementing, following up, and evaluating.

Commercializing involves value transformation to commercial ventures creating end-to-end solutions. Innovation, adoption, and integration require certain intermediate steps including assessing technology readiness, prototyping and testing, risk assessment and mitigation, sustainability planning, manufacturing, financing, marketing, and customer acceptance.

The technology development, adoption, and insertion in a business process or venture need to go through a stepwise road map including:

1. Identifying and/or defining the issue or problem
2. Developing or creating a solution
3. Describing competitive advantages of the proposed solution
4. Selecting a technology or technologies for the proposed solution
5. Assessing its current technology readiness level
6. Developing commercialization plan
7. Arranging appropriate financing
8. Building prototypes for testing
9. Manufacturing or building hardware or system
10. Testing the market and collecting customer feedback

The critical factors for technology commercialization roadmap are prototyping, testing, and identifying the challenges in technology insertion in an existing venture. Creating values for a complete venture depends primarily on financing, marketing, resources, and management capability.

Transformative technology is an emerging technology, and its commercialization along with other emerging technologies can sometimes bring mutual benefits and advantages. When considering integration or insertion of transformative technology into an existing plant, certain digital technologies/techniques (machine learning, digital twins, smart sensors, artificial intelligence, advanced analytics, industrial internet-of-things, and cyber security) become very important.

For technology insertion in an existing manufacturing plant, consideration must be given to industry priorities. These priorities include improving reliability, operability, competitiveness, and profitability. The industries also look to eliminating interruption and wastes. Where applicable, the manufacturing plant would like to adapt to agile methodologies and techniques.

While designing a product, the industry wishes to maximize its benefits from digitizing real-time data and optimizing it for sustainability. The preference is also for obtaining top operational performance and maintaining market leadership. When considering a technology platform, industries usually opt to select Smart Manufacturing technologies. They also prefer to be within a platform that allows them making continuous improvement of their operation.

1.4.1 PLANNING

Financial analysis with justification is the core function of a business plan, and it must include ROI estimate on the capital requested. The justification normally has two parts (1) capital investment requirements including expenses and (2) financial benefits including ROI or pay-back-period. The best-case scenario would show large return with small investment. But in reality, the top strategy would be to select the one with most compelling benefits with good return that requires minimum investment. Consider the measurability of the proposed ROI as the first pick. To increase the probability of success for securing the approval of funding, it is essential to identify the hardware that can be reused in the operation after prototyping.

Considerable numbers of books have been published on developing Business Plans. These books describe on how to develop a complete Business Plan for a venture. A Business Plan can involve details and require considerable time and financial resources to prepare. At the early stage of commercialization of any transformational technology venture, it is very important to start with a brief, but precise, "Summary" of ideas for guidance, reference, and for sharing with others. Such Summary Business Plan should include at least problem identification and definition, solution (product/services) proposed, target market(s) with customers, sales/marketing strategy, business model, identified competitions, competitive advantages, financial projection, resources (technical, management, marketing, financial, regulatory), and target schedule. It is necessary to have a paragraph summarizing the Business Plan. The Summary Business Plan needs to include a section on how you plan to raise capital for supporting the initial years of the commercializing venture.

The planning section in the plan tells us what needs to be done and when, and what benefits they bring for satisfying the overall objectives. The planning document is sometimes called Project Execution Plan (PEP). This planning document will normally consist of the following steps: (1) definition of the tasks to be performed, (2) statements on the objectives of the venture, (3) identification of potential problems, (4) lists of assumptions, (5) definition of the objectives, (6) statements on strategies, (7) identifying sequence of the tasks, (8) establishing required resources, (9) reviewing and finalizing the plan, (10) establishing probability of success, (11) finalizing the plan after a detailed review, and (12) review of the plan often and improve, as background data do change.

It is prudent to prepare an Early Work Schedule. The schedules establish the activities to be performed in sequence. The scoping and tasking prioritization include:

a. Task Identification
b. Logical Sequence of Tasks
c. Allocation of Resources
d. Duration of Each Task
e. Identification of Critical Tasks/Critical Path
f. Leveling of Resources
g. Preparation of Bar Charts

It will be a prudent approach to review the plan at least every 6 months and adjust the plan where appropriate. Situations (particularly market and financing) controlling commercial ventures do change frequently.

1.4.2 FINANCING

During the early stage of technology development and testing, the required funds come from the owner. Also, owner's family and friends are sources of initial financial support. Once the technology has been tested and reports are considered bankable, then borrowing of funds can have little flexibility. Equity crowdfunding is an option, an online offering of new venture's securities to a group of individuals for investment. This option is guided by the country's regulations on financial transactions and securities. The other sources of financing are the conventional private funding individuals. These individuals will ask for significant portion of the assets or shares. Once the prototyping and market testing are done, the corporate or institutional investors would be interested.

To attract investors, it is very important to prepare a believable sales and revenue forecast. Quite often, the owner of the commercializing venture is not equipped to develop a successful financial plan. It may be prudent to retain a Financial Advisor to work with the owner. The Financial Advisors are knowledgeable about investor's preferences in all segments of private and public capital markets.

The feasibility of financing will be based on answering the following questions:

1. What are each shareholder's reasons for participating in raising capital?
2. What share of ownership does each existing shareholder wish to retain?
3. What level of risk will each existing shareholder accept?

4. Is the cash flow forecasted sufficient to justify each shareholder's investment?
5. Will the anticipated cash flow be sufficient to attract capital from outside?

The tax structure(s) in the appropriate country will have a significant impact on attractiveness for the investment.

1.4.3 COMPETITIVENESS

The success of commercializing a venture is primarily based on a transformational technology that has to be monitored, adjusted, secured, and assured often. Successful technology commercialization venture must be sustainable. Critical components in commercialization roadmap includes a minimum technology development, prototype design and testing, financing, manufacturing, regulatory compliance, and customer acceptance. The management team must have adequate knowledge and experience to adjust the action plan as the situation changes. All of these issues must be a part of the sustainability plan.

1.4.4 SUSTAINABILITY

The sustainability plan should address at least the following issues:

1. Technology development
2. Regulatory compliance
3. Testing and manufacturing resources
4. Financing (as/when needed)
5. Market intelligence
6. Management continuity
7. Third-party resources for hire
8. Economic conditions

The sustainability plan should include a "Commercializing Project-Lifecycle Roadmap," identifying the pathways to follow for commercializing a transformational technology. This Lifecycle Roadmap should include the following:

a. Technology/business strategy
b. Feasibility requirements
c. Revenue optimization criteria
d. External resources for validation
e. Permitting guidelines
f. Financing requirements and appropriate sources
g. Constructability analysis
h. Product launching techniques
i. Customer feedback

Required cash flow and expertise are scarce commodities for any venture, especially for first-time ventures. These commodities can be brought in by carefully selecting external resources as partners and by offering them certain rewards or consideration of values. This may include sharing a part ownership (thus sharing the risks as well) of the venture and/or sharing profits at negotiated terms and conditions acceptable to both parties. The approach could include the following:

1. **Strategy and Objective:** The partnering objectives and strategies are created during the Business Plan development. As the roadmap is developed and the proposed venture starts taking shape, the partnering strategy starts taking shape. Overall, objective(s) may remain the same, but implementation strategies can change.
2. **Needs Assessment:** Needs assessment should continue at different stages of the roadmap, as the situation(s) do change every so often.
3. **Selection of Expertise:** During each phase of product development, it is expected to face different challenges. So, the need for different types of expertise will change at different stages of the roadmap. The planning process should be flexible enough to recognize and secure expertise as required.
4. **Hire or Subcontract:** Most of the time, the needs for additional expertise will be for a short time. That dictates whether the venture should consider hiring or subcontracting. Consider hiring when the need is continuous and for longer time. Subcontracting is preferable when the need is for a shorter period or as and when required.
5. **Technical Collaboration:** When certain activities of a product venture are not required on a continuous basis, then it may be a prudent practice to consider having technical collaboration with another organization with expertise and adequate resources. Sometimes, collaboration with an organization having market recognition for technical expertise and cost-effectiveness should be considered highly desirable.
6. **Board of Directors:** Sometimes, the members on the Board of Directors are selected because of their expertise. These directors' technical, business, or financial expertise can be great assets, at a minimum cost to the venture.
7. **Support Services (Short-Term vs. Long-Term):**
   a. **Individuals vs. Organization:** Quite often, product venture struggles with the benefit assessment on having an individual employee or engaging an organization as a subcontractor. The benefits on both options can be difficult to access.
   b. **Accounting and Financial:** Most often, it is prudent to engage an outside accounting firm for Accounting and Financial services.
   c. **Legal and Intellectual Property:** Also, the product venture (or start-up companies) retains contracted services for Legal and IP support. This approach can be quite cost-effective.
   d. **Manufacturing:** During the early stage of the venture, the sales are not strong enough to require own manufacturing facility. It may be a prudent approach to use external manufacturing facility during the early life of the venture. This approach may be costlier than having your own facility. But strategically, the external manufacturing option is a lot more desirable considering financial commitment and expenditures for having your own facility.

    e. **Quality Control:** It will be a prudent practice to use external, reputable QA/QC services. A reputable third-party QC Report will carry much value with the investors or the client community.

    f. **Specialty R&D Support:** Only in special cases, consideration can be given to using an external Specialty R&D Support who has reputation in solving a specific R&D issue.

Creating Sustainable Values for products and for establishing sustainable business venture support is essential in sustainable commercialization venture. Here, expert knowledge is essential. The skills needed are:

    a. Entrepreneurial leadership to building commercial ventures
    b. Teaming for sustainability
    c. Customer care essentials

The business ventures need to have effective end-to-end solutions for sustainability and successful commercial operation. Creating values and developing competitive advantages for sustaining commercial venture are essential in a competitive environment. In technically dominated market sectors, resources for nontechnical aspects are usually forgotten or marginalized. Discussions on certain components of commercialization roadmap can be nontechnical but are essential for any successful commercialization.

Inability to understand the customer requirements can also be a cause of business venture failures. Depending on the market segment, a marketing specialist with appropriate experience should be hired. The experts are not available all the time, especially when you need them.

For creating or supporting a successful venture involving technology commercialization, adequate planning with details is essential for the total roadmap. Lot of time, it is easy to miss the required tasks in a critical stage of the venture. This detailed plan is very important to assess project status at any time, specifically for understanding when to change the course of actions, if warranted. The detailed plan is required for each phase so that one could have the total control of the program.

For improving the success factor (probability), it is suggested that a Program Manager with appropriate experience needs to be considered for the success of the commercialization venture.

1.4.5 CUSTOMER ALLIANCE

The Product Launching Plan (PLP) is the last and final plan for the commercialization venture. After successful implementation of all preceding action items in the commercialization roadmap, if the PLP is not developed properly, the commercialization venture most likely will not be successful. The PLP should cover at least the following steps:

1. **Objective:** The purpose of the PLP phase is to introduce the manufactured product to the potential customers and collect their feedback on presentation, packaging, pricing, usability, desirability as repeat buyer, and suggestions for improvement, if any.

2. **Planning:** Much consideration, time, and resources are to be devoted for planning the PLP.
3. **Schedule (Phasing):** The PLP efforts should be scheduled and, if necessary, should be phased for convenience.
4. **Control Criteria:** Identify the specific data that must be collected or gathered for control. Characteristics of control issues must be defined and planned from the beginning.
5. **Staffing:** Identify the key staff and their activities and allocate as appropriate.
6. **Documentation:** Develop the forms including data collection formats.
7. **Customer Feedback:** The customer feedback should be recorded in pre-designed forms.
8. **Budget:** Allocation of financial resources should be set before the PLP phase.
9. **Reporting:** Identify the key officials that should review the data collected and provide their comments for next action(s).
10. **Authority:** It is very important to identify the officials who have the authority to review and approve the PLP. The authority of officials to be designated, for any changes.

Business success is created through the eyes of the customers. Proper Customer care helps create better revenue and profit performances. Customer experience can influence the perception of the product and viability of the organization. Many organizations have developed several techniques on how to improve the customer perception of the product, its pricing, competitiveness, and sustainability. A good strategy is to collect customer feedback, understand the issues, and create the program that will not only maintain the level of confidence of the customers but also improve on it gradually or step-wise. The program must address both tactical and strategic issues.

In the eyes of the customers, pricing of the product is very important. This involves affordability and competitiveness. One of many challenges is to evaluate the pricing of the competition, keeping a lookout for promotional programs of the competition. Based on market intelligence, the timing and content of the promotional programs should be developed and implemented. Also, it is important to understand the impact of your promotional activities.

It is essential to understand why customers buy products at a price they are willing to pay. So, it is very important to get customers' feedback, so that immediate actions can be taken for any negative comments. It is important to develop a system for engaging with the customers regularly. Customers can also tell us the negative aspects of your competitors' products. So, it is absolutely essential to design the customer care program with utmost care. The commercializing roadmap also needs to develop policies on pricing and promotion including guidelines on customer feedback.

1.5 MARKET TRENDS AND BROADER OPPORTUNITIES

The world of manufacturing is vast. One requires viewing the broader market potentials beyond the current interests that exist in the technology community. Certain applications of technologies may be overlooked initially, but these will be tagged once their sustainable benefits are recognized. One can visualize the technologies that can currently impact our everyday life. But there are applications with huge potentials and a small push is needed to make them into commercially available products. Currently, some of these ideas are confined in academic research but could become commercial products in the near future. The opportunities are to be created or developed under some of the following transformational domains.

1.5.1 ARTIFICIAL INTELLIGENCE/MACHINE LEARNING

AI is being applied in the market place to secure greater benefits. AI in association with machine learning is now a powerful tool for the industries. AI along with IIoT helped develop "Industry 4.0" platform. These innovative technologies are helping manufacturing plants become more efficient and competitive, especially in the oil and gas, chemical, petrochemicals, mining, and manufacturing sectors.

Adoption of AI with nanotechnology will have wide applications. There are certain developing areas where AI converges with nanotechnology. In the imaging area, AI benefits scanning probe microscopy. Applications in this area have resolution issues particularly in imaging samples at nanoscale. This approach can help to make a more efficient imaging system. From a simulation perspective, there are many different parameters that need to be correlated to accurately produce either an image or a moving depiction of a working system. AI can better collate the data, learn from past systems, and produce a more accurate representation of the system. AI can adopt nanocomputing. The computation is performed through nanoscale devices and a wide range of AI methodologies can be applied to nanosensors.

1.5.2 DIGITAL TRANSFORMATION

Our world has been going through digital transformation for quite some time. By harnessing digital data, certain products can take its place in our social and economic domains. We harness the nano-enabled data in real time so that we can manage smartly our health, energy, food, transport, infrastructure, water resources, education, security, communications, and finances. We can expect to see many nanotechnology-based products entering the smart living domain in this digital age.

1.5.3 INNOVATIVE INFRASTRUCTURES

With expanding urban areas, construction is a vital part of its growth, particularly for infrastructures. Building materials can provide a positive impact in the composites used in building materials. The concrete and cement are using certain materials. Some materials have high tensile strengths and can be used in stress and strain gauges during and after construction to ensure that the buildings are structurally sound. Technology can also enable managing extreme heat, cold, and degradation. Currently, road signs are storing data and helping create greater safety and security for smart cities. In addition, latex inks can be used in road signs to track traffic movements improving the transportation security. Materials are used in eco-paints, making it anti-corrosive coatings. Eco-paints can also remove pollutants from atmosphere. Graphene has many applications, including as an additive to asphalts for improving the road's resistance to wear. Special materials are already being used to make buildings more flexible and resistant to harsh environments and pollution. Special technology-enabled buildings will make it easier to be built with better resistance to the worse perils. These materials are already being used to make buildings more flexible and resistant to harsh environments and pollution, including removing carbon dioxide from atmosphere. Innovative technology-enabled buildings will make it easier to build with better resistance to the worse perils of earthquakes or even tsunamis in years to come. Future cities will continue to grow for accommodating increase in populations and for durability and sustainability.

1.5.4 GLOBAL WARMING

Severe climatic changes are serious global challenges, such as huge number of hurricanes, flooding, and heat waves. These cause extreme casualties. Reversing the effects of severe climatic change is very difficult but can be managed through reducing greenhouse gas emissions, primarily carbon dioxide released to atmosphere. For over a century, fossil fuels are being used to generate power and these produce a large quantity of greenhouse gases with carbon dioxide emissions causing global warming. During the last couple of decades, fossil fuels are being slowly replaced by renewable energy such as solar and wind energy. In this energy transition, nanotechnology can come to rescue.

While going through this transition, we are still faced with continued use of polluting fossil fuels for many years to come. So, it makes sense to have technological solution for removing carbon dioxide entering the atmosphere. This can be done through carbon capture technology using nano-sized membrane structures. Carbon capture is the process of removing carbon dioxide emissions after fossil fuels are burned. Carbon capture technology relies on the principles that certain gaseous molecules are captured by an appropriate membrane. Membranes specially developed for the purpose can capture up to 90% of the carbon dioxide released in the combustion process. This can cause a significant reduction in greenhouse gas emissions to the atmosphere. The membranes are reusable as the pores can be cleaned of any carbon deposits without damaging.

1.5.5 SMART CITIES

Major cities around the world are growing fast. They are growing to accommodate ever-increasing populations and many of their support systems are becoming highly automated. Certain transformational technologies are already impacting many aspects of city life, and their applications will increase in the future. Cities are applying new technology to enhance the capabilities of their existing infrastructures. Smart cities will enable complete technological transformation for enhancing their existing platforms. With new materials, such as graphene, the industries are developing special sensors and graphene batteries. This will help the optimum usage of time and energy. Smart cities are already utilizing transformational technologies in many of their systems including Internet of Things (IoT), big data, and machine-learning algorithms, enhancing existing infrastructure performances. Smart cities are going to see the application of advanced sensors providing extensive localized data points. Nanotechnology enables tiny sensors to manage extraordinary amounts of data across multiple platforms. Certain smart cities in the world are currently implementing 5G technology in their cities and are busy in laying super fiber-optic cables so that more data can be transmitted through these 5G networks. It will provide support for remote environmental monitoring applications, personalized medicine, and early intervention in healthcare systems. The revolution in healthcare will bring advanced point-of-care diagnostic tools at home allowing accurate self-diagnosis in real time using highly advanced mobile devices.

1.5.6 EMERGING MARKETS

In recent years, opportunities have opened up in the emerging markets based on work done primarily in the nanotechnology sector. Some of these potential opportunities are presented here:

Building Materials: Many materials including graphenes have been in use in the construction industry.

Anti-Corrosive: Use of new materials such as nano-titania, nano-silica, nano-zinc-oxide, and nano-silver in certain special paints enhancing anti-corrosion properties.

Carbon Nanofibers: Carbon nanofibers are being used for reinforcing concrete roads in the snowy areas.

Explosives: Powerful explosives have been developed with nanomaterials. Coated nanotubes provide a burst of power to the smallest systems.

Heat Insulation: Nanoparticles are being used in the heat-insulating surface layers that protect aircraft engines from heat. It increases the service life of the coating by 300%.

Heat Sink: New carbon nanotube sheets are demonstrating world's top heat-sink performance.

Defense: A limited number of initiatives have been undertaken to develop nanoproducts for application in the defense area. Some of these potentials are:

Advanced Camouflage: Work has progressed in testing nano-biomimicry applications in advanced camouflage and intelligent uniform.

Explosive Detection: Certain nanotechnology can soon replace bomb-sniffing dogs. Tests have identified a way to increase the sensitivity of a light-based plasmon sensor to detect minute concentration of explosives.

Performance Enhancement: Thin film nanotechnology is being tested for soldier's performance enhancement and safety.

Soldier Protection: A nanomaterial-integrated textile is being tested for enhancing survival capability of soldiers in extreme environments.

Warmer Clothing: Fabrics embedded with nanowires and hydrogels can help soldiers stay warm and comfortable in colder climate.

Weight Reduction: Printed nanoscale structures are being considered for weight reductions in defense systems.

Electronics: Several initiatives have been undertaken to develop nanoproducts in the electronics area. Some of these potential commercial applications are:

Biochips: Nanoscale 3D Printing Technique is being used for micro-pyramids to build better biochips.

Broadband Light-Detector: Tests have found a way to control how the material conducts electricity by using extremely short light pulses, enabling its use as a broadband light detector.

Data Storage: Nanomaterial "Graphene" pushes the possibility closer to data storage devices.

Edible-Electronics: Electronics that dissolve or even be edible. This could be the answer to the ever-growing problem of e-wastes.

Electronic-Structure: Tests using a beam of ultraviolet light can look deep into the electronic structure of a material made of alternate layer of graphene and calcium.

Electronic-Devices: Flexible electronic devices such as e-readers could be folded to fit into a pocket. This involves designing circuits based on carbon nanotubes (CNTs) in place of rigid silicon chips.

Magnetic Circuits: Nanoscale magnetic circuits are found to expand into three dimensions.

Magnetic Graphene: Successful tests have shown that induced magnetism in graphene while preserving graphene's electronic properties.

Nanoribbons: Graphene nanoribbons can be applied in molecular levels. Graphene combined with amorphous carbon increases its signal transmission efficiency and stability for use in semiconductor devices.

Energy: Energy market has investigated reasonably well in finding appropriate applications of nanotechnology products in energy. The opportunities include:

Batteries: The application of nanomaterials is starting to gain traction in the field of batteries. While the detail investigations are yet to be carried out to assess the long-term safety, efficiency, and charge/discharge cycle rates, the potentials look very promising for many applications. The lithium battery with silver vanadium diphosphate electrodes has high potential. Sodium's abundance and low cost gives it an advantage over lithium for energy storage. Tests have found a way to make sodium-ion battery practical. An electrode nanomaterial shortens the diffusion distance of sodium ions, giving batteries much better rate performance than their predecessors. Graphenes in batteries are examples of potential commercial application of nanomaterials. But it will take time for testing the requirements in meeting regulations at least for safety and health. In addition, sheets of graphene, graphene balls, carbon nano-scrolls, silver nanowires, and various lithium-based thin films could be used as electrode matrix. Many of these electrodes developed are known to store a larger amount of charge, have more efficient cycling rates and greater overall efficiencies compared to current graphitic based electrodes.

Energy Storage: Nanomaterials will have wide use as the energy storage mediums in everyday electronics. Utilization of nanomaterials is considered safe over long time periods. Mobile energy storage technology has started using special graphene material to boost significantly the energy density of electrochemical batteries.

Solar Cell: Researchers have developed a comprehensive model explaining how electrons flow inside new types of solar cell. The model allows better understanding of such cells and helps in increasing their efficiency. In solar cells, some of the light energy is lost as heat. Recent work has paired up polymers that recover some of the lost energy by producing two electrical charge carriers per unit of light instead of the usual one.

Solar cells rely on semiconducting junctions to convert the solar energy into electricity. Many nanomaterials have been widely established as materials that can be used in these junctions. Currently, there are many different types of solar cell that employ nanomaterials because they provide greater conversion efficiency over traditional solar cells. Traditional inorganic solar cells composed of silicon and indium-tin-oxide (ITO) are slowly being replaced by nanomaterials such as graphene, quantum dots, and 1D nanowires.

Sustainable Grid: The insulation plastic used in high-voltage cables can withstand over 25% higher voltage if nanometer-sized carbon balls are added. This will assist in improving efficiency gain in the power grids for a sustainable energy system.

Transistors: Three potentials for nanomaterial use in electronics are being considered.

a. **Bilayer Graphene:** Tests have been conducted successfully on the behavior of bilayer graphene to verify whether it could replace silicon transistors in electric circuits.

b. **Engineered Graphene:** Research work indicates that the engineered bandgap brings graphene near to displacing silicon.

c. **NanoRing:** NanoRing Transistor has high promise for future applications in electronics.

Engineered Materials: Nanomaterial "graphene" has many targeted applications, and has much more potential. Graphene is a two-dimensional material and is known for its versatility. Here are some of the targeted applications.

Graphene: Graphene has excellent physical properties that made a significant impact in the construction industry. These properties include electrical conductivity, tensile strength, and optical transparency. Graphene is used in composites and as an additive to cement for high-strength, self-cleaning, and flexural-strength. It also helps making the cement environment friendly.

Electronics Application: Scientists used doped graphene that allows addition or subtraction of electrons from it by chemical means. The experiment revealed that a doped graphene absorbs a single photon and it excites several electrons proportionally to the degree of doping. The photon excites an electron that rapidly falls back down to its ground state of energy. As it falls, it excites two more electrons on average as a knock-on effect. A photovoltaic device using doped graphene could show significant improvements in device performances.

Graphene Membranes: Graphene is hydrophobic as well as hydrophilic. It is stronger than steel, flexible, bendable, and 1 million times thinner than a strand of human hair. Previously, graphene oxide membranes were shown to be completely impermeable to all solvents except for water. Tests have shown that one can tailor the molecules that pass through these membranes by simply making them ultrathin. Newly developed ultrathin membranes are assembled in such a way that pinholes formed during the assembly are interconnected by graphene nanochannels and they produce an atomic-scale sieve allowing the large flow of solvents through the membrane. This greatly helps the applications of graphene-based membranes for sea water desalination and organic solvent nanofiltration (OSN).

Graphene Nanoribbons: A new study shows elegant mathematical solution for understanding the flow of electrons changes when CNTs turn into zigzag nanoribbons.

Graphene Foam: Graphene foam combined with epoxy is substantially tougher than pure epoxy and is far more conductive than other epoxy composites. Epoxy-filler brings better conductivity at the cost of weight and compressive strength, but the composite becomes harder to process. It replaces metal with a three-dimensional foam made of nanoscale sheets of graphene. Easy interlocking between the graphene and epoxy helps stabilize the structure of graphene.

Brittle Smartphone Screens: Scientists have developed a new way to make smartphone touch screens that are cheaper, less brittle, and more environmentally friendly. The new approach promises devices that use less energy and do not tarnish in the air. Currently, indium tin oxide is used to make smartphone screens and it is brittle and expensive. The primary constituent, indium, is also a rare metal and is ecologically damaging to extract. Silver is the best alternative to indium, but it is expensive. Combining silver nanowires with graphene enabled a two-dimensional carbon material. The new hybrid material matches the performance of the existing technologies at a fraction of the cost.

Carbon-Nanotube Catalysts: Researchers used a new class of single atom catalysts (SACs) supported on CNTs that exhibit outstanding electrochemical reduction of CO_2 to CO. Nickel single atom nitrogen doped CNTs (NiSA-N-CNTs) are considered to have the highest metal loading for SACs. Single atoms of nickel, cobalt, and iron were supported on nitrogen-doped CNTs for comparison.

Diamond-Hard Materials: Researchers have developed a material that is flexible and lightweight as foil but is stiff and hard enough to stop a bullet on impact. Flexible and layered sheets of graphene temporarily become harder than diamond and impenetrable upon impact. By applying pressure at the nanoscale with an indenter to two layers of graphene, researchers transformed the honeycombed graphene into a diamond-like material at room temperature. Graphite and diamonds are both made entirely of carbon, but the atoms are arranged differently in each material, giving them distinct properties such as hardness, flexibility, and conduction of electricity. The new technique allows manipulation of graphite so that it can have beneficial properties of a diamond under specific conditions.

Silver Nanowires: Silver nanowires have been used in touchscreens before, but it is the first time it got combined with graphene. Combining silver nanowires and graphene on a large scale is easier by spraying machines and patterned rollers. The addition of graphene to the silver nanowire network also increases its ability to conduct electricity.

Bottom of Stronger-Silk: Spider silk is one of the toughest polymer fibers known. Its properties are enhanced even further if the spiders are allowed to ingest graphene and CNTs. The enhanced fibers could find uses in specialty high-performance fabrics, such as parachutes, or in biodegradable applications such as in sutures or medical dressings. Spider silk is stronger than conventional silkworm silk and therefore has received much research attention in recent years. It is among the best spun polymer fibers in terms of tensile strength, ultimate strain, and toughness, even when compared to synthetic fibers.

Surface Texturing: Nanostructured surface texture, with shape inspired by the structure of moth eyes, would prevent the reflection of light off silicon that would improve the conversion of sunlight to electricity. Etching a nanoscale texture on to the silicon materials creates an antireflective surface working as thin-film multilayer coating.

Touchscreens: Graphene has an excellent potential for use in touchscreens on phones, tablets, laptops, watches, and interactive whiteboards. Flexible touchscreens will help enhance its use for wearable electronics/wearable technology. Graphene's optical transparency and flexibility helps its high-tech applications. Graphene is slowly entering into sportswear that is durable, thermoregulating, and lightweight.

Manufacturing: Breakthroughs of innovative application in industrial sectors with nanomaterials include in manufacturing also. Benefits come primarily from creating materials that exhibit certain properties, such as stronger, lighter, and better conducting of electricity. These can help create new manufacturing processes including for additive manufacturing. The most promise for innovative manufacturing technologies is based on new hybrid materials and nanotechnology that allow the generation of very small features in very large surface areas.

Microscale Assembly: Currently, there is no way to assemble components across multiple-size scales. Very tiny features of nanotechnologies can enable assembly of components that is very difficult to hold in a device. Researchers are using modulated surface energy to control the adhesion of flexible tools to manipulate assembly. Tests show potential in developing a set of tools with varying sizes of different soft contact tabs that can pick up very small devices and then reposition them. The traditional method is to place the devices in very high-end equipment where tiny robotic arms do the assembly, one by one.

Photolithography: Semiconductors can use high-density small patterns on polymers in microelectronics processing. Current photolithography methods are extremely expensive but effective. Researchers are working on an optical process that avoids the expensive methods available. The process works by partially using the material's response to light while also controlling the light interacting with the material. By combining these, one can create very sharp features that are very small. It is low-cost, scalable, and once developed can have a large impact on how electronic components are made and how much they cost.

Robotics: Robotic systems have led to mass assembly of nanostructures.

Semiconductors: Semiconductors are very important for the rapidly growing fields of consumer-and-industrial electronics and optical sensing. Both miniaturization and high-volume processing are important in producing affordable, yet complex circuits used in many devices.

Water Supply and Sanitation: Innovation has transformed conventional sanitation systems for more than 2 billion people in the world not having access to adequate sanitation. Innovation is going to bring greater sustainable benefits in the sanitation area globally. Another area where technological advancement includes creating nitrogen compounds in water. This is normally achieved through biological conversion, but this is a slow process. Application of palladium to catalyze the conversion of nitrate to nitrogen speeds up the process enormously. This reaction suffers from the drawback that it produces the harmful by-product ammonia. Top of Form Researchers have assembled 2D materials with sub-nanometer slits that hold potential for water desalination. These materials are made from graphene, hexagonal boron nitride (hBN), and molybdenum disulfide (MoS2).

2 Operational Technologies

Scott M. Shemwell

2.1 INTRODUCTION

"To me, ideas are worth nothing unless executed. They are just a multiplier. Execution is worth millions."

–Steve Jobs

There is a large body of information regarding Smart Technologies and Smart Manufacturing. We will draw on some as cited herein. This chapter is all about execution. "I want a Smart Manufacturing company—How do I get one?" We will show you how!

"Smart Manufacturing aka *Smart Factory/Industry 4.0/Manufacturing 4.0* is the notion of orchestrating physical and digital processes within factories and across other supply chain functions to optimize current and future supply and demand requirements. This is accomplished by transforming and improving ways in which people, process, and technology operate to deliver the critical information needed to impact decision quality, efficiency, cost and agility."[1] It is also referred to as the 4th Industrial Revolution (4IR).

In this section, we will address Smart technology–enabled Operations Models, such as the Internet of Things (IoT) depicted in the following figure. In addition to a focus on the promise of using business cases in addition to the theoretical construct, the full life cycle and associated process will be included.

As with any new initiative, risks must be assessed and understood and a mitigation strategy developed and implemented. Moreover, the Convergence of Information Technologies with Operational Technologies must account for the Human. We will also look at the role of Human Factors as this process unfolds.

The final output of this section is an actionable Roadmap for transforming your production facilities into Smart Manufacturing. The Roadmap is just that, a set of guidelines, milestones, and a destination complete with alternative routes and a risk profile for each.

2.1.1 ONE CAVEAT

A Glossary and other supporting documents as well as a list of key resources that can assist you on this journey are also provided. This section takes the reader through the complexity of transforming the firm in a simplified manner without losing the fidelity demanded by the new business model.

One of the challenges of researching this broad topic is the availability of unbiased data and information. There is a wealth of information and many case studies are available. However, much of it is in the form of marketing materials from equipment

DOI: 10.1201/9781003156970-2

and software providers as well as service implementation providers of all sizes. We have tried to remove bias as much as possible and acknowledge it when we can.

Therefore, in some cases, we reference sources such as Wikipedia which may contain secondary data and information. However, the Wisdom of Crowds construct supports the hypothesis that collectively the wisdom of many is greater than a single expert.[2] We believe this holds true for the robust and complex subject.

2.1.2   KEY IMPLEMENTATION ACTION ITEMS

The information technology (IT) advisory firm, Gartner®, has put forth a six-step smart manufacturing strategy.[3] It is important to understand that with any transformation effort, the organization and its ecosystems must be able to accomplish the tasks in a timely manner. In other words, there should be actionable, realistic items that can facilitate issue resolution.

The Gartner Six Action Items include:

- **Putting People First:** For decades change management processes have acknowledged that individual stakeholder in a transformation process must come first. This must include decision-making process, risk mitigation strategies and well as the set of organizational cultures and governance across the ecosystem.
- **The Integration of Continuous Innovation with Continuous Improvement:** Aligning today's technology and transformation initiatives allowing for future solutions and process change. Historically, the term often used was "Future Proofing."
- **Performance Management – Efficiency vs. Speed:** Proponents of Digitalization have long posited that speed or time to solution implementation is the goal even at the expense of Operational Excellence. Readers should take caution with this approach. Rapid implementation is very critical but at what price?
- **Execute Gradually:** Decades of IT deployments have shown levels of organizational disruption, especially with the initial rollout and pilot. Each plant may have different operational business drivers and even seasonal product, i.e., gasoline blends. Avoid the 'cookie cutter' approach. The dichotomy— digitalization is about speed; yet enterprise-wide deployment takes time.
- **Realism:** Develop Use Cases that rank order value as well as risks. In other words, take the 'low hanging fruit' first to show early successes. However, understand the overall value vs. risk and deploy the solution accordingly.
- **Roadmap:** The development of a strategic, realistic project plan, budget and risk profile is critical for success. It should also address resource required as well as the technology assessment process.

For more than 2 years, it has been very challenging for organizations with global operations dependent upon complex supply chains. Beginning with the emergence of the Covid-19 global pandemic in late 2019, further complicated by global/ regional politics and issues such as the Texas electric power grid failure in February 2021, the grounding of the container ship, the Ever Given, blocking the Suez Canal in late March of that year are forcing operations management to scramble.[4]

Undoubtedly, during the period from this writing to the publication of this book, many more Critical Path issues will negatively impact Operational Excellence initiatives. Moreover, the level of complexity continues to grow with an expected massive increase in the amount of data supporting decision-makers (human and machine)—cyber security will remain a top priority. Finally, societal demands are increasing especially on heavy industry sectors such as energy and petrochemicals.

2.1.3 MANAGING COMPLEXITY

We humans can deal with only a certain number of variables or action items at one time. However, many claim the ability to multitask, and some are proud of their belief in superhuman powers to deal with complex problems even in our head without writing it down.

There is also a tendency to overly simplify complexity when presenting to management and others. Classic examples include the three PowerPoint bullets and the Green—Yellow—Red colored risk management matrix.

Many readers are familiar with data reduction techniques used to help manage large volumes of data in an ordered and categorize in a taxonomy. In this chapter, we will provide tools to help advisors and decision-makers deal with both large numbers of numerical variables as well as often conflicting business rules and processes.

Complex and evolving systems are difficult to manage by definition. However, making incorrect and poorly thought out business and technical models can and have led to disasters. Not just plant failures and accidents but more frequently technology implementation and transformation initiatives.

Decision support models must be understandable by all involved, yet they cannot lose the fidelity required because of their complexity. Moreover, they must be timely in their response to inputs.

2.1.4 RAPID RESPONSE MANAGEMENT

In 2009, the author put forth a construct that rapidly evolved into the approach toward *agility*, *resiliency*, and *sustainability* necessary for successful Smart Manufacturers of any size or geographic disposition. We recognized that there are basically four categories and chose to use a sports metaphor.

There are two basic components. First the 'level' of play or capability as evidenced whether an organization can be categorized as High School Varsity, Weekend Athlete, Master Athlete or Olympian. For the sake of argument, Collegiate and Professional-/Semiprofessional-level teams should fall under the Olympian classification. The other variable is 'Organizational VO_2max' a useful KPI for measuring athleticism—agility/sustainability.

A high-level taxonomy is depicted in this figure. It is important to understand that it is satisfactory to be in any group and moving to the upper right-hand quadrant is not necessary. Just recognize the value and limitations for the sector chosen. Moreover, moving to any category has a financial and organizational cost (including cultural changes) so management must assess the value proposition for any transformation.[5]

The message herein; not every firm will be an Olympian, nor should they be—any quadrant can be fine. Like any new technology, Smart Manufacturing needs to have a fit-for-purpose component.

It is important to understand that 'one size does not fit all' and may actually be detrimental if the appropriate assessment process is not followed. This will be further addressed in this chapter.

2.1.5   GUIDE TO THIS SECTION

There is a wealth of data and information positing new and better ways to lead operations. Much of it assumes that coupling IT with Operational Technologies will lead to superior performance. Would that be true?

During these next segments, we will review and assess current thought leadership subjects and seek to provide readers with an actionable and realistic plan to add stakeholder value across the entire Environmental Social and Governance (ESG) criterion set. Readers can expect to take useful action items that can be used immediately in their daily business.

High School Varsity	Olympian
This organization is often new; start up or spin out firms. Filled with youthful enthusiasm it challenges the status quo. Occasionally, they burn brightly and then plunge to earth in a meteoric fireball. • Often outliers, initially in the 5 percentile • Early stock growth can be explosive, and they often far outperform in the near term • Kudos from all stakeholder quarters • Organizational VO$_2$max is Very Good	This organization is at the top of its game, with strong endurance, rapid recovery from adversity, and seemingly flouting market forces. • Consistently a top performer across business cycles. • High long-term stock multiples • Focus of envy and fear from competitors • Organizational VO$_2$max is Superb
Weekend Athlete	**Master Athlete**
This organization is in pretty good shape. We define this as the situation most competitive firms will find themselves. They are slower to recover than other athletic firms. They are often the protectors of the status quo, resisting change as not applicable to them. • Generally, they fall well within two standard deviations of industry metrics • Stock performance is middle of the road in up markets and falls more in down ones • Often, they are viewed as good but not great by analysts • Organizational VO$_2$max is Average	This organization is at the top of a mature game. Maturing from the High School Varsity or throttling down from the Olympian, they dominate their niche (perhaps in a thin or mature segment). • Consistently a better than average performer but not stellar • Better than average performance in the stock market • Believed by many to be a good solid player • Organizational VO$_2$max is Good

FIGURE 2.1   Rapid response matrix.

Given the massive transformation underway, it is likely that the Austrian economist Joseph Schumpeter's Creative Destruction is at work. Likely, this will result in a hybrid yet significantly different Operations World on the backside of this process.

Then, it will happen all over again! Particularly as the *Velocity of Information* accelerates.[6] Our goal is to provide the reader with a structured actionable Roadmap that can be successfully implemented immediately and sustained.

2.2 INFORMATION TECHNOLOGY OPERATIONAL TECHNOLOGY

Readers may be familiar with the emerging construct whereby IT and Operations Technology (OT) converge. For decades, the argument has raged that IT did not understand the business and business leaders did not understand IT.

Often, this was an excuse for the poor performance of field Operations Management solutions. No doubt there was some truth in this belief. For example, in the writer's experience, no operations person wanted to be assigned to an Enterprise Resource Planning (ERP) project as it was seen as possibly career ending. Some of these biases may remain in some firms and is a component of project risk.

In the era of Management Information Systems (MIS) this was the likely scenario for all but forward-looking firms. Ever in those days' organizations that had a good understanding of their business value using data and information held strategic advantage over their peers and competitors.

Today and going forward, the gap between these two technology suits must be bridged if manufacturing firms are to survive much less thrive. The model put forth in the following figure adds the Human element as the glue that seals this gap.

As shown in the figure, both technology suites reflect typical and even classic tools. However, in both cases, as has been shown in the earlier chapter, current and emerging IT enables manufacturing firms to implement the IoT aka Industrial Internet of Things (IIoT) business model as supported by sensor and other field, shop and equipment technologies facilitating smart solutions.

As will be discussed later in more detail, advances in business models, such as High Reliability Management (HRM) and Human Factors among others, can take full advantage of IT-OT. However, briefly we can define the Human Element three components herein:

- **Human Factors:** "The application of psychological and physiological principles to the engineering and design of products, processes, and systems."[7]
- **High Reliability Organization (HRO):** "Use systems thinking to evaluate and design for safety, but they are keenly aware that safety is an emergent, rather than a static, property. New threats to safety continuously emerge, uncertainty is endemic, and no two accidents are exactly alike."[8]
- **Human Systems Integration (HSI):** "A comprehensive, interdisciplinary management and technical approach applied to system development and integration as part of a wider systems engineering process to ensure that human performance is optimized to increase total system performance and minimize total system ownership costs."[9]

The point is that without the Human Element, typical IT-OT models are only partial. It is the human aspect that completes the implementation solution.

Moreover, IT-OT must be aligned with the business model, governance construct, and must add economic value to the organization and its various stakeholders. Finally, it must be built on manufacturing and IT Standards as well as Generally Accepted Business Practices including Supply Chain Management (SCM) Good Practices.

There are several points about Operational Technology (OT) that need to be taken into consideration during the transformation processes that are probably well known by operations-oriented individuals but may not be as well understood by management and those on the IT side. For example:

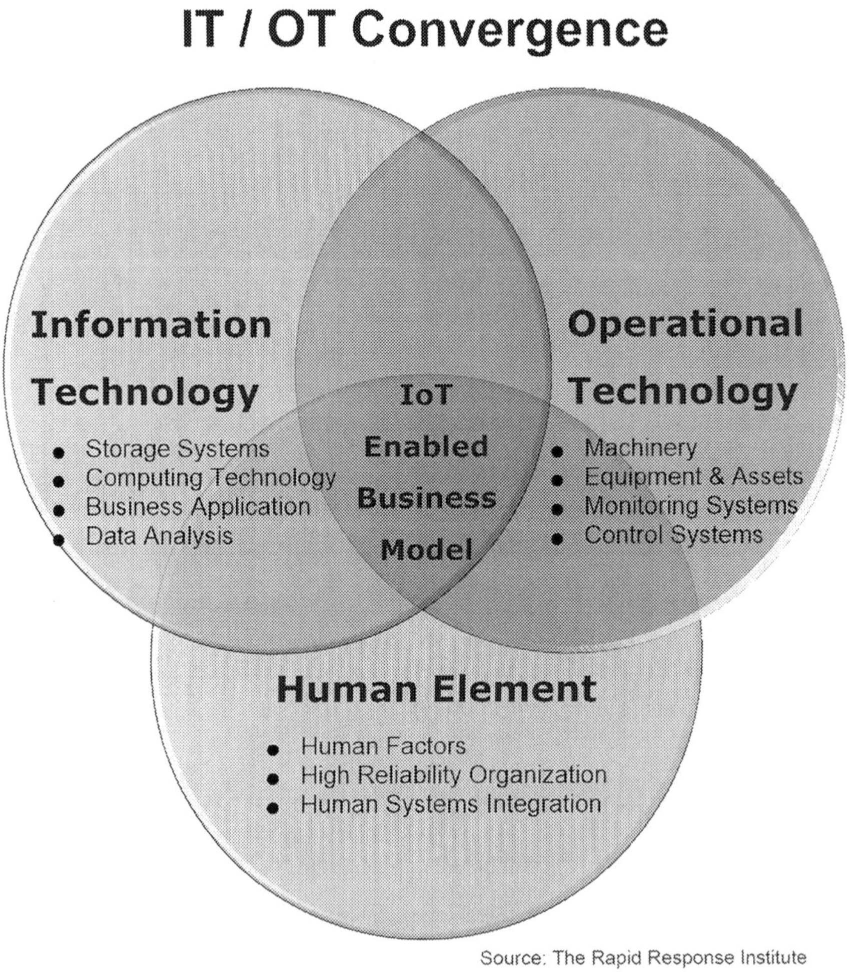

IT / OT Convergence

Source: The Rapid Response Institute

FIGURE 2.2   IoT-enabled business model.

- OT systems are usually legacy systems, in some cases decades old. As such they may have vulnerabilities to such things as cyberattacks and must have new levels of resiliency build into them.
- They also can take decades to update and upgrade and when these processes are undertaken, systems may have to go offline, i.e., production maybe shut down perhaps for several days.
- Up to 60% of this workforce may need to be *reskilled* or *upskilled*. For example, incorporating cybersecurity into traditional system reliability and safety.[10]

These and other issues will be addressed in this chapter. It is important to realize that the existing enterprise-level systems are a function of their lifecycle, and each will have its own set of criteria to address during this transformation process.

This basic Smart Manufacturing construct steers all aspects of the implementation of robust solutions driving better manufacturing economics and values. We will build on this throughout this chapter.

2.2.1 Human vs. Machine

We are at the cusp of dramatic technological capabilities that can transform organizations, industries, and our own careers. However, there is a cautionary tale herein. How do we know the output from these decision systems is correct, i.e., Valid and Reliable?

Moreover, there is a tendency to put too much faith in technology. A recent example of misplaced trust and poor human interface is the Boeing 737 Max 8.[11] We are also glued to our smart phones and tablets. Interestingly, this phenomenon is not new.

Circa 1975, while a 27+ year old senior field engineer logging oil and gas wells in Louisiana this author was training a younger junior engineer. On one calculation using a then state-of-the-art HP engineering calculator, the junior engineer arrived at the incorrect answer.

It was immediately clear that the calculator 'spit' out the wrong result. More importantly, the junior engineer was not aware that his answer was an error!

Trained on the slide rule, a quick mental calculation quickly told this more senior observer that the output of the calculator did not reflect the correct answer. Quickly assessing the order of magnitude and two significant digits, it was clear the 'machine' had made an error.

No doubt the algorithms within the calculator were probably correct. After all, this calculation was relatively simple and straightforward.

The logical culprit was human error, most likely an input typo. When asked about this clear problem, the junior engineer simply stared at the machine as if it had 'betrayed' him.

Why does this 45-year-old story matter? The junior engineer simply made a mistake, and the more seasoned individual caught it, and it was corrected. No harm no foul, right?

In this case, no lives were at risk and the error would have eventually been corrected during an audit process. This simple example continues. The junior engineer did not have and/or understood the 'context' of his decision.

His IT system provided him a number; however, he was not able to assess its accuracy. Did it solve the problem? Had his solution stood, an error would have been introduced into the process.

He singularly believed in the output of a computing system, aka the DATA. Flash forward, 'it's all about the data stupid,' as the saying goes. Other more critical examples include aircraft autopilots where pilots either do not believe the output or do not understand, aka Situational Awareness.[12]

In 1999, the $125 million Mars Climate Orbiter was suddenly lost. NASA discovered that legions of US-based engineers had missed the fact that certain software used the 'Imperial' US measurement standard instead of the 'Metric' system. A high school physics problem.[13]

Issues surrounding data acquisition and assessment will be further detailed in this chapter. This area is one of the most critical risk aspects of Smart Manufacturing and more often than not, the human is the primary culprit in data integrity issues.

2.2.2   HUMAN FACTORS ENGINEERING

The human–machine interface is as old as the development of the first tool. In today's sophisticated environment, there are two major issues that must be addressed. First, the system must be set up (dashboard/other) so that the human operator can properly assess conditions and make decisions. Second, the system must provide enough insight into the data so that the human can override as appropriate.

The second condition is the most difficult and requires significant training as well as input from the digital system to alert the human that the system is being overwhelmed. One example of an Alarm Management problem is when failures in one or more sensors or processes start providing operators with inconsistent, misleading, and even wrong information—Alarm Overload.[14]

It is helpful to define several overlapping terms. Three will be described, the following two and HSI in the next section.

- The International Ergonomics Association defines *Ergonomics or Human Factors* as, "the scientific discipline concerned with the understanding of interactions among humans and other elements of a system, and the profession that applies theory, principles, data, and methods to design in order to optimize human well-being and overall system performance."[15]
- *Human Factors Engineering* is also defined as ergonomics refers to a body of knowledge regarding human interface with equipment of all types. It also refers to the process of designing machines that address the safety, productivity, and comfort of the human operator. Moreover, multiple disciplines are integrated to accomplish these goals.[16]

Moreover, governance and management must develop and foster a Safety Culture that empowers individuals to make decisions without personal ramifications

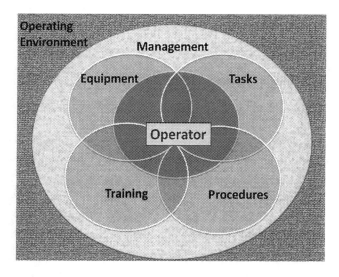

FIGURE 2.3   Human factors operating environment.

if the human action is wrong. Both HRM and a Safety Culture demand this empowerment.

This figure depicts the human operator at the center of the operations universe. This person is responsible for assuring high performance, and safe and effective operations.

Management must set the cultural stage for human factors to add value to an organization and its ecosystem. In the complex Smart Manufacturing environment, human factors will be more important than ever.

2.2.3   HUMAN SYSTEMS INTEGRATION

"Human Systems Integration (HSI) is the management and technical discipline of planning, enabling, coordinating, and optimizing all human-related considerations during system design, development, test, production, use and disposal of systems, subsystems, equipment and facilities."[17] The overarching model of the human—machine interface for complex systems.

Originally developed by the US Department of Defense (DoD), this model has found broad take-up across many industry sectors. There are generally seven domains:

1. **Human Resources:** The portfolio of Knowledge, Skills and Abilities (KSAs) necessary to perform the tasks properly and adequately.
2. **Personnel:** Typically, the softer skills necessary to manage complex systems as opposed to skill sets mentioned above.
3. **Training:** Cost-effective solutions and tools (including eLearning) necessary to 'reskill' or 'upskill' the workforce and sustain current and future requirements.

4. **Human Factors Engineering:** Defined earlier.
5. **Safety and Occupational Health:** The requirements to assure safe working conditions.
6. **Production and Survivability:** Those factors necessary to assure reliability, sustainability, cyber security, and HRM as required.
7. **Habitability:** Satisfactory working and if required living conditions. I would add Safety Culture to the category as well.

There is a great deal of material available on HSI. Standards, metrics, and performance models are readily available and some or more may be useful/required in Smart Manufacturing initiatives. However, it is beyond the scope of this chapter to perform a 'deep-dive' into the subject.[18]

2.2.4 HIGH RELIABILITY MANAGEMENT

Conventional wisdom has been that complex systems will inevitably fail, and the negative impact is unavoidable with limited recourse. Normal Accident Theory (NAT), posited by Charles Perrow circa 1999, remains the classic approach to the management of interrelated systems theoretical without empirical evidence.[19] This construct is widely believed as evidenced by two Space Shuttle disasters, Deepwater Horizon, Bhopal, and other major highly visible industrial disasters.

However, NAT is not necessarily the only way to manage a business. High Reliability Theory (HRT) puts forth the hypotheses of focus on failure instead of success, and the tactical aspects of current operations as well as an emphasis on resilience.[20]

The following 'process' definitions are taken from our book, 'Implementing a Culture of Safety: A Roadmap to Performance-Based Compliance,' wherein we posited that High Reliability played a major role in transforming the oil and gas sector to a Safety Culture–driven model. The definitions of the five processes that follow are taken from that volume.[21]

Preoccupation with Failure: HROs treat any lapse as a symptom that something may be wrong with the system, something that could have severe consequences if several separate small errors happened to coincide. They also make a continuing effort to articulate mistakes they don't want to make and assess the likelihood that strategies increase the risk of triggering these mistakes.

Reluctance to Simplify: There is often a desire to simplify a complex situation by reducing options to "High, Expected, or Low." Management, it is said, "does not have time or interest for greater details." HRT suggests that trying to simplify a process that is complex by nature risks creating more risk, not less.

Focus on Operations: For most energy firms, the largest revenue stream and greatest exposure to shareholder value is field operations. Operational Excellence only happens when top management makes it the priority. HRT

confirms the importance of these business and technical processes to overall organizational health.

Capabilities for Resilience: NASA demonstrated the ability to respond to
unforeseen and certainly unexpected conditions during the Apollo 13 flight.

Unstructured Organizational Structure: In other words, provide personnel
with the flexibility to respond to unplanned events. This is directly contrary
to NAT processes many companies now follow. Empowering individuals
has been a subject of discussion for decades. Now may be the time to actually do it.

The following figure depicts the role these five processes have in setting Mindfulness
toward management of unexpected events driving toward greater or higher reliability. One can also view this framework from the perspective of rapid response.
When failures occur knowledgeable reaction time is of the essence. HRM enables
this capability.

In its 2019 article, "What high-reliability organizations get right," the consulting firm, McKinsey & Company described three core business processes driving
reliability. First, their research indicated that human dynamics, rigorous processes,
clarity of roles, and accountability matter just as much as technology. They go on to
point out that successful HROs put a premium in certain skills and invest in training
accordingly. One can also make the case that the 2021 ransomware cyberattack of the
Colonial Pipeline suggests that HRM techniques could help remedy the increasing
onslaught of cyber criminality.

We believe HRM has a major role to play in Smart Manufacturing. Therefore, it
should be part of any transformation undertaken.

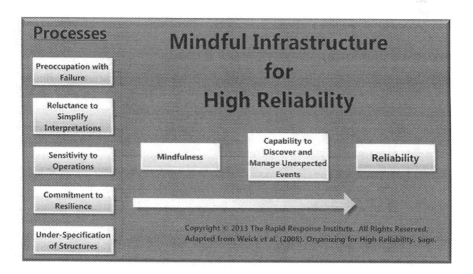

FIGURE 2.4 High reliability mindful infrastructure.

2.3   VALUE PROPOSITION

This section develops the construct for assuring Smart Manufacturing initiatives actually add value to the firm. Later, we will document economic models to help assess the value and guide the project processes.

"The digital business case must be aligned with the organization's long-term business priorities e.g., growth projections, talent retention and sustainability goals."[22]

Return on Investment (ROI) for IT projects have been elusive. Moreover, many careers and not just IT executives have floundered following project rollout. A 2017 study by the Project Management Institute (PMI) indicated that 14% of IT projects are 'total' failures. In addition, 31% did not meet goals, 43% were over budget, and 49% were late.

While there will always be a level of uncertainty and risk associated with any project, especially large ones, PMI compiled the following reasons why IT projects are at risk of failure[23]:

- **Inaccurate and/or Requirements:** Poorly developed and understood Requirement Documentation can lead to many difficulties especially in today's video meeting situations and multiple remote teams. Thirty-nine percent failures are because of poor requirements documents.
- **Poor Project Executive Sponsor Involvement:** Often executive sponsors are simply appointed and may or may not have active involvement. Especially if projects are struggling or he or she does not understand it well, there may be some political expedience taken. This accounts for 27% of failures.
- **Shifting Objectives:** Projects often suffer from a large number of Change Orders. There may be valid reasons such as equipment delays, but often they result from poor requirements documentation. Project management techniques, such as Waterfall, Agile, and Scrum, attempt to mitigate changes in scope, time, or budget. Changes herein account for a 36% failure rate.
- **Inaccurate Estimation:** At the beginning of a project, there is a high level of uncertainty. This study suggests that poor estimating practices and planning account for failure in 28% of IT projects.
- **Unexpected Risks:** Risk mitigation is key to all phases of the Smart Manufacturing lifecycle not just the project stage. This phase has a lot of unknowns and uncertainties, so the risk is high. Therefore, simple risk models are no longer acceptable in complex industries. Unexpected risk cause 27% of failures.
- **Dependency Delays:** With large complex projects like Smart Manufacturing, it is likely that a large number internal and supply chain (including subcontractors) will be involved. This can create dependencies that can negatively impact on one or more of project tasks—contributing to 23% of failures. Techniques such as the Critical Path Method (CPM) can be useful planning tools to determine which tasks may impact the overall project time and how to manage their schedule.[24]

- **Resource Constraints:** The availability of individuals with the necessary KSAs is a major constraint and is often not satisfactory to demand everyone give 125% or more with stretch targets. This very true for large long-term projects. Human loading is a key CPM constraint accounting for approximately 22% failure rate.
- **Poor Project Management:** This is an obvious issue and is not limited to the Project Manager (PM), but others such as Schedulers and others. This can also include operation's personnel assigned and accounts for 20% of failures. Projects are not static plans and Gantt Charts but dynamic entities requiring aggressive, productive, knowledgeable, and high-performance management.
- **Team Procrastination:** Late deliverables even at the plan development stage can begin to derail a project quickly and it may never recover. This human weakness must be dealt with as soon as teams show signs of it—11% of failures

This overview of poor project execution is cursory at best. Interested readers should consult additional resources on this subject. However, there are two mitigation factors completely within control of the firm.

A strong and committed Executive Sponsor needs to be in place with specific performance incentives. In addition, a strong and experienced PM and direct report staff is a fundamental necessity. Finally, the Roadmap can provide additional guidance in addition to robust project management processes.

These numbers and issues are consistent with the author's 40 plus years perspective as a team member or Subject Matter Expert (SME), Program manager, Executive Sponsor, and Customer. The challenges are real and should not be underestimated.

2.3.1 Expected Value of Marginal Information

Whenever a statistician uses the term *expected*, he or she is referring to the outcome of a stochastic model. In other words, the probability or likelihood of an event or issue occurring.

Many models are deterministic in nature. Their focus is on a single outcome. For example, a traffic light is red, yellow, or green but not any combination thereof.

Stochastic models on the other hand reflect any number of possible outcomes. For example, the probability of winning the lottery may be one in a million or more.

Today, effective risk mitigation models are stochastic in nature. The older risk models that reflect the traffic light mentality do not reflect reality and needlessly exposes the organization. New models are constantly updated with operational data both from the firm as well as exogenous or external data. A simple example is the weather forecast developed based on third-party or government data.

Investment in technology, especially IT, always needs a business or Use Case. Often investments in new and unknown fields such as Artificial Intelligence (AI) are difficult to measure. There are many models available that address economic value assessment.

However, a simple construct based on Economic Utility Theory has been available for approximately 25 years. While the details of this model are beyond the scope of this book, a robust well documented model is available for interested readers.[25]

Briefly, Economic Profit (EP) is a function of the Return on Invested Capital (ROIC) and the Weighted Average Cost of Capital (WACC). Typically, heavy industry (and investors) measures performance as a function of ROIC.

Therefore, it makes sense that ALL Capital Expenditures (CAPEX) be treated equally. If for no other reason, then the executive management sees the investment world through this lens.

What follows is:

The probability (stochastic) that new information attained at no economic costs between different options exceeds the probability of decisions made based on existing information is greater than zero, then the investment is warranted.

An NPV in excess of the marginal utility of information represents economic value to the firm.

All business and use case assessment models should be based on these economics. If they do not pass this test, then their outcome is suspect.

The problem does not lie in the model. The difficulties with *validity* and *reliability* are from the data set and the algorithm used to populate this construct.

2.3.2 MAXIMIZING CAPITAL EFFICIENCY

How does an organization achieve sustained competitive advantage when all organizations have access to the same relatively inexpensive technologies, i.e., the marginal cost of technology is effectively zero?

Pretty simple, it is a function of the organization's culture and its ability to use technology better than its competitors. A simple example, Sears Roebuck had the same business model as Amazon.

Allow customers to select from a catalogue, order, pay, and wait for delivery. Sears was a major success story for decades. What happened? Perhaps, complacency and bureaucracy? If it can happen here, can't it happen anywhere? The answer is yes.

Revenue, costs management, and metrics toward bonuses are not the answer. Surely Sears employed those Key Performance Indicators (KPI).

Maximizing Capital Efficiency is the root metric Boards demand. One wonders how well they understand this KPI?

Typically, capital efficiency can be assessed from Return on Capital Employed (ROIC) previously mentioned as a function of EP. It is the core metric used by many firms and bankers/investors.

Note as a general rule, management is only looking at the economic metrics they can control. For example, the 'total life cycle costs' of a capital good, i.e., an automobile, are not taken into consideration. Manufacturers cannot impact the development of raw material such as steel from mining, only the price they pay for materials. This issue is a major source of discussion in the renewable energy sector.[26]

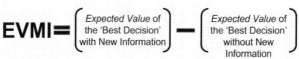

Economic Value of Marginal Information

$$EVMI = \begin{pmatrix} \textit{Expected Value} \text{ of} \\ \text{the 'Best Decision'} \\ \text{with New Information} \end{pmatrix} - \begin{pmatrix} \textit{Expected Value} \text{ of} \\ \text{the 'Best Decision'} \\ \text{without New} \\ \text{Information} \end{pmatrix}$$

New information is defined as having no economic cost

© Scott M. Shemwell

FIGURE 2.5   Economic value of marginal information.

2.3.3   Conclusion

Developing a meaningful and documented value proposition is perhaps the most important aspect of any new digitalization initiative. This is a particularly difficult task because there are so many unknowns or 'squishy' variables to consider.

In the remainder of this chapter, we will build on this construct to develop a useful Economic Value Proposition Matrix (EVPM) that directly addresses hard cost, i.e., cost of a new facility as well as the softer intangibles such as productivity enhancements.

2.4   ENVIRONMENTAL, SOCIAL, AND GOVERNANCE

The use of advanced operational technologies coupled with process reinvention will enable global firms and their partners to capture new value using sophisticated ESG models.

Investopedia defines ESG Criteria as, "a set of standards for a company's operations that socially conscious investors use to screen potential investments."

- *Environmental* criteria consider how a company performs as a steward of nature.
- *Social* criteria examine how it manages relationships with employees, suppliers, customers, and the communities where it operates.
- *Governance* deals with a company's leadership, executive pay, audits, internal controls, and shareholder rights.[27]

The term is relatively new, yet it builds on policies and processes of the past, like all management theory.[28] For example, the term governance is rooted in Agency Theory—The principle that seeks to resolve differences between corporate owners (shareholders) and their employed agents (executives).[29]

One can posit that organizations have broad responsibilities to all their stakeholders: Shareholders, employees, local communities, regulatory agencies, etc. This encompasses the stewardship of the environment and society.

It is important that Smart Manufacturing initiatives align with the organization's culture, mission and vision. If Smart becomes 'the way we run the business,' it must be aligned with ESG as well.

2.5   MATURITY MODELS

Business and technology maturity models are qualitative assessments of an organization's ability to continuously improve in the field or area a particular model supports. Their general approach is usually a five-step process from ad hoc management to structured innovation processes. Typically, there are two implementation approaches: top-down where observed behaviors are 'fit' into one of the steps and bottom up whereby clusters of behaviors generally fit a stage.[30]

Various maturity models have evolved and cover a wide range of processes. Moreover, since the fundamental model has been published in the 1980s, there are many software solution providers and professional services readily available offering implementation and other maturity-oriented services. As with any procurement assessment process, buyers should use the technology assessment process described in later sections if they choose to use these third-party services.

However, it is important to view any maturity model as a 'go by' and not a set of 'tick in the box' processes that must be strictly adhered to. Moreover, in an ecosystem, such as a firm, which would find itself implementing Smart Manufacturing, one must assess the maturity state of suppliers, subcontractors, and even customers when deciding their appropriate level.

Maturity models are not to seek perfection but to understand the appropriate level. Most models advise that to attain levels four and five will require a level of effort and funding that may be high. As with any new business models, the question must be what is the value of the said implementation?

2.5.1   SMART MANUFACTURING MATURITY

It is tempting to address the maturity of an organization's ability to implement Smart Manufacturing as a simplistic straightforward model of several (typically five) steps. However, an initiative this large will have several maturity models that must be assessed, and action plan developed for each is yet in sync.

It is a mistake to assess the maturity of an organization to implement such a broad transformation. Rather it is more about the sum of all the 'maturity' parts that makes the Smart Manufacturing whole. In other words, implementation processes must understand the relative maturities of IT and OT as well as the back office as well as the supply chain. 'Getting out ahead of these skis' can be disastrous.

As with most, if not all, the tools and techniques described herein, there is no one size fits all for maturity models. Moreover, an organization's maturity is usually the lowest level attained by any major group or division in the organization. For example, if 24% of the plants are deemed to be at level 2 and the rest at level 3, the overall firm should have a maturity level of two. This is not necessarily negative, but an acceptance of the reality of the situation.

The overall Smart Manufacturing maturity of an organization is a function of all 'relevant' business and technical process maturities. As stated, generally the lowest maturity level of any set of maturities is the overall level of maturity. Seeking to drive higher without acknowledging this reality can lead to failure and even catastrophic development and ongoing performance.

The next paragraphs describe a set of high-level maturity models that are relevant to Smart Manufacturing solutions. They are not meant to be all inclusive, only to document the major fields of interest. Other resources are available as shown in the Commercial Tools and Services table that follows.

2.5.2 SOFTWARE AND DATA

In the mid-1980s, the Carnegie Mellon Software Engineering Institute (SEI) developed its Capability Maturity Model (CMM) in response to the US government's software development contract's requirements. Later, it morphed to the CMM Integration (CMMI) now administered by the CMMI Institute, a subsidiary of ISACA.

Effectively, a set of Best Practices, CMMI is a very mature model, which consists of:

- The Model
- Adoption Guidance
- Systems and Tools
- Training and Certification
- Appraisal Method.[31]

This is a tool for benchmarking and helping drive the IT implementation process. As a very mature construct, the model has been adapted to industry-specific issues as well as Data Management Maturity. Briefly, the construct consists of these five steps:

1. **Initial (Chaotic, Ad hoc, Individual Heroics):** The starting point for use of a new or undocumented repeat process.

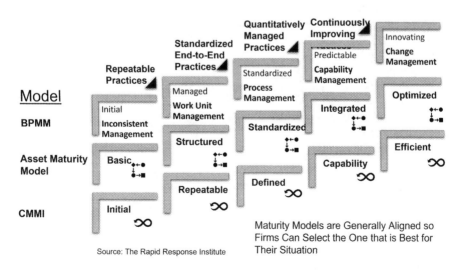

Source: The Rapid Response Institute

FIGURE 2.6 The five levels of process maturity.

2. **Repeatable:** The process is at least documented sufficiently such that repeating the same steps may be attempted.
3. **Defined:** The process is defined/confirmed as a standard business process.
4. **Capable:** The process is quantitatively managed in accordance with agreed-upon metrics.
5. **Efficient:** Process management includes deliberate process optimization/improvement.[32]

Additional information is available from a variety of sources and as stated, there are organizations that can help with the use of CMMI to help implement Smart Manufacturing. Some are referenced in this document; see the tables in the appendices.

2.5.3 DIGITALIZATION

There are a number of commercial Digitalization Maturity Models available.[33] However, most appear to be more market oriented than process refinement or technology adoption solutions. Therefore, they may not be appropriate when assessing where the organization is and what a go forward plan looks like. Interested readers can research and make their own decision.

From the perspective of Smart Manufacturing, the CMMI model is a good 'go by.' We will discuss the 'hype cycle' later; however, much of what is available to organizations without consulting engagement may be considered hype.

2.5.4 PROCESS

A logical follow on to CMMI is the Business Process Maturity. Historically, the view is that IT enables business processes which are the value to an organization's performance. This remains true.

A set of Standards as well as a Maturity has been developed in an attempt to provide significant value to vertical industries, including manufacturing. Similar to CMMI, the Process Maturity Model has five steps as well as insight into change. For example, in the following figure, as a process moves from step 1 or Initial, it adapts Repeatable Processes to attain a Managed State (step 2).[34]

Finally, Process Assessment and Improvement is an ongoing or continuous activity. By one model, four steps have been identified:

1. Prior to any change, the program must clearly identify and outline the outcomes of any change in business processes. Note the plural as any process is a function of subprocesses. For example, even filing an expense account has several subtasks—providing receipts, documenting the reason, completing the form, etc.
2. Setting the metrics and KPIs of success. How will data be collected and normalized, analyzed, and presented to management?
3. Decision-making based on the outcomes and data attained.
4. Adjust and improve accordingly.[35]

Business processes will continue to evolve and even make step-level changes. More mature organizations will most likely be more successful at change and perhaps more importantly be easier to transform.

2.5.5 GOVERNANCE

For decades, organizational governance has focused on financial transparency and assurance against corporate malfeasance. More recently, the emphasis has been more inclusive, i.e., ESG.

Following the Deepwater Horizon incident of April 2010, the San Bruno Pipeline Explosion in September 2010 and the Fukushima Daiichi Nuclear incident in March 2011, the author and his colleagues started assessing why such 'god like' destruction could occur with modern industrial activities and how to prevent future incidents. In addition to certain professional services and even software (see Commercial Tools and Services table), we developed two new business models; Asset/Equipment Integrity Governance (AEIG) and the Safety Culture Implementation Roadmap (addressed in the next section).

The AEIG is based on the following four pillars:[36]

- **Maturity Level:** Similar to the CMMI model.
- **Portfolio Management:** The understanding that any organization is a function of its portfolio of revenue-producing assets.
- **Policies and Procedures:** Foundation and featured in the Operations Management System (OMS) depicted herein.
- **Criticality:** Focus on those issues/variables that have the highest and greatest impact. Rank ordered risk mitigation.

Preceding ESG by a decade, this model demands management pay attention to those processes that can kill people, destroy production, and devastate the environment. It is appropriate to incorporate this model into Smart Manufacturing decision-making processes to assure that this does not happen in the future.

The Three Mile Island nuclear accident in 1979 provided the foundation for the Culture of Safety requirements. Many sectors have implemented self-imposing tenets. The emerging recognition that systematic safety is more than OSHA requirements such as hardhats while important, are lacking.

While they vary between sectors, the following tenets effectively capture the intent of Safety Culture driven firms. Safety Management Systems (SEMS) are typically built on the following tenets and designed to reduce incidents as well as fatalities.[37] They incorporate the ISO 4500 family of Occupational Health and Safety standards but are performance based in nature, meaning that *Systematic Safety* is behavioral economics by nature and a KPI.

- **Leadership:** Focus and commitment by top-level management along with Board oversight.
- **Hazard Identification, Risk Management, and Resolution:** Clear and open processes to find and fix issues.
- **Personal Accountability:** Empowering at all levels to take responsibility for actions.

- **Safe Work Processes:** Tasks and other activities have safety integrated seamlessly into the workflow.
- **Continuous Learning:** Formalized ongoing reskilled or upskilled training the workforce
- **Foster an Environment for Raising Concerns:** Allowing 'anyone' regardless of their role to bring up concerns without recrimination.
- **Effective Communication:** Focus on safety and HRO to assure success without incident.
- **Trust and Respect:** A respectful work environment.
- **Inquiring Attitude:** Fostering individual right to engage with their work and identify issues of concern.

These tenets are all about maximizing the safe performance of the Human Capital in the ecosystem. It is important to understand that a Safety Culture is not separate from Operations Management. It must be embedded with the work processes.

Personnel cannot be expected to perform tasks and then fill out a safety form. The process must be all inclusive in OMSs.

An organization's culture must be one of Safety Culture not an add-on to the 'way we do business.' Inherent safe processes and HRO must be endemic to our business model.

2.5.6 CONCLUSION

Maturity models are proliferating and often presented as standalone or in an organizational vacuum. While it is important to assess and understand where the organization fits within these boundaries, they are only guidelines and must be put into the context of the overall organization and even the industry sector.

Later in this book, we briefly refer to 'The Talent Development Maturity Model.' This model will serve the same purpose to help the firm reskill and upskill its workforce. It will also apply to senior executives as well. Expect other maturity models that will need to be incorporated into the overall firm's maturity.

Moreover, with so many maturity models, no firm will be 'best-in-class for all.' Management must decide how any given model is aligned with the business. Finally, if one views the set of maturity models from a portfolio perspective, some will be more critical than others. Effectively, it is a weighted average approach.

2.6 TECHNOLOGY TAKE-UP

When something new surfaces, there are several responses. Some may 'jump' on the idea, and some may see it with some trepidation. The process of assimilation varies among groups.

It is important for management to understand how and even if the organization will accept and internalize new technologies with their process changes such as required by Smart Manufacturing initiatives. There are several models available that can help management visualize the process.

However, like most issues Smart Manufacturing must address, processes may not be clear and in some cases are controversial. Adopting a given methodology is part of the project risk management process organizations must decide upon.

We address two accepted models in the next paragraphs. Neither is sacrosanct but each described a scenario that may be helpful. Readers can decide the best path for their organization.

2.6.1 Adoption Model

There are standard technology and process adoption models; some have been around for several decades. They can be useful from the perspective of the maturity and risk appetite organizations and even sectors have about their willingness to bring new processes and technologies into the organization.[38]

Generally, there are five levels or rates of adoption[39]:

1. **Innovators:** Those who like the 'new' and are the first to try technology despite the higher risk associated with untested products and/or solutions. Less than 3% of organizations are in this category.
2. **Early Adopters:** While not as quick to take up new technologies, these individuals/organizations are willing to invest time and funds ahead of others. Approximately 13.5%.
3. **Early Majorities:** Comfortable when they see others having success, this 34% moves at a slower rate.
4. **Late Majorities:** These skeptics are slower on the uptake and represent approximately 34% of the population.
5. **Laggards:** Very risk adverse 16% and respond only after technology is mature, proven, and widely available.

As noted, the risk profile lowers with each step with typical expectations that waiting can mitigate exposure. In addition, the figures below show a range represented by the gap in the 'S curve' which is labeled Population Density to reflect a level of uncertainty of the take up rate.

Moreover, 'Critical Mass' is attained when the adoption rate becomes self-sustaining. Value continues to grow both for the technology provider as well as the growing user base.

One way to determine "expected" or statistically driven value using this model is to look at the area under the curve. We can actually calculate the expected value using Integral Calculus. This is shown in both figures as a function of the series of (bars) rectangles.[40]

If one views the traditional adoption curve, it appears that the greatest expected value could be derived by waiting and letting others take the risks associated with new technologies. However, this would not capture the highest value.

By looking at the adoption lifecycle from the perspective of the following figure, adjusted for risk, the greatest value is derived by obtaining and using technology much earlier. Perhaps, Innovation is too early for enterprise-level use. Early Adoption or even Early Majority may be more appropriate, and a large amount of value can be taken.[41]

This traditional model can help organizations see where they might fit vis-à-vis others. Most firms in the Critical Infrastructure sectors tend to wait for technologies to mature. The risks of the unproven can be high in certain applications and culturally most wait. However, it may be time to change that mindset when it comes to Smart Manufacturing solutions.

One can make a case that Amazon incorporated new technology early in its lifecycle while Sears did not. If this hypothesis is supported, this could be a contributing factor to the demise of the original mail order retailer founded in 1893.

2.6.2 Gartner

There are a number of major IT advisory firms, many have been acknowledged in this book. One of the more visible is the publicly traded Gartner, Inc. The firm is quite large with revenues in excess of US$ 4 billion.[42]

In addition to a host of services, the firm has developed two tools that are of direct interest to this subject matter and in many cases are free to use. These include:

- **Magic Quadrant:** A graphical perspective of the relevant standing of key technology providers categorized as Leaders, Visionaries, Niche Players, and Challengers.[43]
- **Hype Cycle:** Assessment of the maturity and viability of new technology as a function of their value toward solving business problems.[44]

The firm has other tools and solutions, and interest parties should investigate further. Moreover, as mentioned, there are more tools and documentation available from other Gartner partners as well that might be of use too.

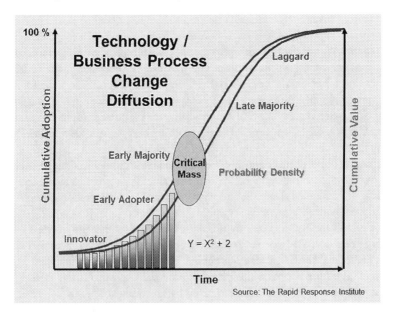

FIGURE 2.7   Traditional adoption model.

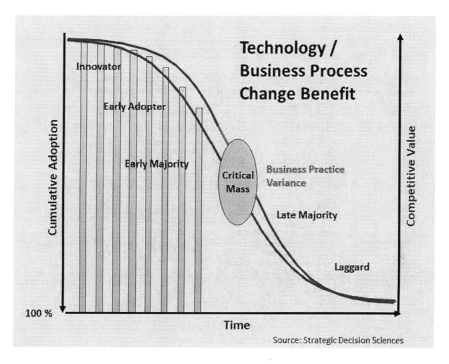

FIGURE 2.8 Transformed adoption model.

2.6.3 MAGIC QUADRANT

The firm has developed a set of Magic Quadrants that can help decision-makers. As with a number of business tools, the upper right-hand quadrant is where companies would like to find themselves.

This graphic representation of the competitive landscape is comprised for sectors:

- **Leaders:** Driving their current capabilities and well positioned for the future.
- **Visionaries:** Know where markets are heading but need more work in the execution process.
- **Niche Players:** As the name implies focused on a smaller segment.
- **Challengers:** Execute well but lack an understanding of market direction.

The following figure is a generic view of this tool and gives readers an idea of how it can be used to quickly assess relative competitive positioning.[45] It is pretty straightforward qualitative and subjective perspective on market segments.

One recent example important to the manufacturing sector is the Magic Quadrant for Robotic Process Automation released in July 2021.[46] The 2022 version of this graphic is shown below. The firm provides online access to a large knowledge base of Magic Quadrants and Critical Capabilities. This information is readily available.[47]

The Magic Quadrant

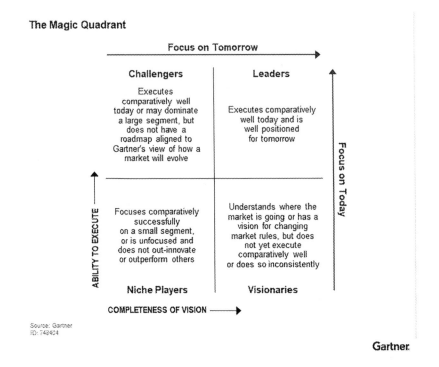

FIGURE 2.9   Gartner magic quadrant. (Gartner, How Markets and Vendors Are Evaluated in Gartner Magic Quadrants, Cecile Drew, Julie Thomas et al., 20 April 2021. *GARTNER is a registered trademark and service mark of Gartner, Inc. and/or its affiliates in the U.S. and internationally and is used herein with permission. All rights reserved.*)

One should keep in mind that graphic tools of this nature provide a high level and superficial perspective. They should not be relied on exclusively and pre-procurement research must dive deeper into stated positions.

2.6.3.1   Hype-Cycle

The title is telling. We live in an era of exploding technologies. Daily, some new whiz-bang widget is announced and—Hyped! No longer better than sliced bread, new technology promise everything from global peace to weight loss.

Marketers have long 'puffed' products, i.e., this is the best car in the world, best tasting hamburger, etc. Subjective statement by 'pitch men/women.' Nothing wrong or illegal with these statements, but there is a lot of commercial noise to sort through.

A major challenge for those making the serious investment in digitalization solutions such as Smart Manufacturing is to sort through the noise and assess the 'current' value and RISK of new technologies to the organization's mix.

The tool consists of five phases:

- **Technology or Innovation Trigger:** Potential breakthrough but unproven.
- **Peak of Inflated Expectations:** Early and always positive publicity. The innovator and early adopters may take first mover advantage, but the risk can be very high.

- **Trough of Disillusionment:** Oops it is not working out and reality sets in. Later adopters become more cautious.
- **Slope of Enlightenment:** A more mature product/solution starts to show promise and better understood. Some remain cautious.
- **Plateau of Productivity:** The technology finally arrives and take up is broad based.[48]

As we will show, these models go into more detail and the use of this tool requires that users go into much greater detail. As with the Magic Quadrant, the firm has identified a large number of discrete areas to investigate.

While this tool is useful, it should not be relied upon exclusively but used in conjunction with other robust risk assessment processes. Moreover, it is not dynamic but updated periodically so newer solutions may not be reflected in documentation.

Finally, we will provide an actual relevant example in the Roadmap section that follows. The results might surprise some readers.

2.7   REFERENCE ARCHITECTURE

This particular discussion is on the technical side and hence will only be addressed at a high level. However, every major manufacturer's IT department has a reference architecture, which is key to successful implementation and use for 'all' IT initiatives regardless of application.

Effectively, it is a software template that provides a common set of standards including a common vocabulary (see Center or Excellence). For manufacturing, ISA-95, International Standard for the Integration of Enterprise and Control Systems, may be an applicable standard for most organizations. This appears to be an emerging standard and based on the circa 1990s Purdue Reference Model for Computer Integrated Manufacturing (CIM), so the relevance of this hierarchical model may be dated.[49]

That said, this legacy system structure is in place for many organizations and will need to be incorporated into Smart Manufacturing at least during the transition to newer models. Some advocate a hybrid model. One that capitalizes on the *status quo* software composition, incorporating the next step without sacrificing safety and operational performance. Such a hybrid solution integrates IT and OT data flows from the horizontal Industrial IoT (IIoT) model with the traditional hierarchical structure.[50]

As part of the transformation process, it is necessary to discover and have a plan for the existing revenue-producing IT assets. Other Reference Architectures exist, and the above discussion is not meant to preclude them. Appropriate Due Diligence is critical and must involve the general IT Enterprise Architecture and how Smart Manufacturing would fit as well as the long-term (years) transition process.

Finally, and as shown in the following figure, much of the technology/process issues that must be addressed in the Enterprise Smart Manufacturing initiatives are relatively immature. For example, all the following are no further than the Trough of Dissolution in the hype cycle:

- Agile Beyond IT
- Agile Project Management
- Composable Applications

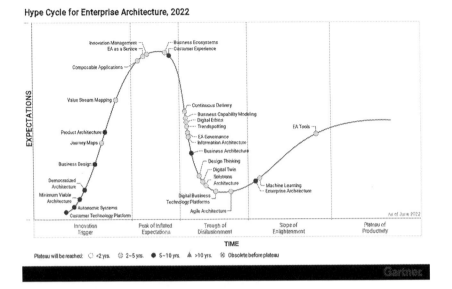

FIGURE 2.10 Hype cycle for enterprise architecture, 2022. (Gartner, Hype Cycle™ for Enterprise Architecture, 2022, Saul Brand, Marcus Blosch, 20 June 2022. Gartner and Hype Cycle are registered trademarks of Gartner, Inc. and/or its affiliates in the U.S. and internationally and are used herein with permission. All rights reserved.)

- Digital Business Technology Platform
- Digital Twin
- EA Governance
- Innovation Management
- Machine Learning

This explicitly suggests that risk profile is higher than more mature technology and processes. Not necessarily negative but 'just the way it is' and must be identified and appropriately addressed.

The Enterprise Architecture appears to be the most mature.[51] It may make sense to build Smart technology solutions around them—probably easier to define a valid value proposition to management.

2.7.1 Conclusion

It is important to understand that many of the challenges faced when implementing Smart Manufacturing may result from the immaturity of solutions available. This is not to say as has been previously stated that these projects are not undertaken only that the risk profile may be higher and should addressed accordingly.

2.8 PROCESS INTEGRATION AND OPTIMIZATION

Organizations are dynamic and undergo continuous metamorphose. Hopefully, continuous improvement—if not ushering in demise or acquisition. Therefore, firms strive to attain and sustain Operational Excellence.

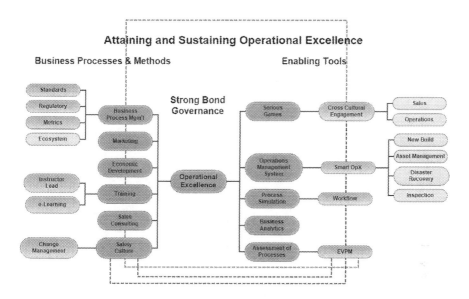

FIGURE 2.11 Operational excellence framework.

A robust Operational Excellence Platform has been developed and was originally published by the author in 2018.[52] A brief overview of this model is appropriate as Operational Excellence and Smart Manufacturing are symbiotic.

This platform is straightforward. Operational Excellence depends on three things. First a strong governance model underpins the entire construct. The other two components are Processes and Methods as well as Enabling Tools. Notice that it is built on our Relationships, Behaviors, Conditions (RBC) Framework presented later.

2.8.1 PROCESSES AND METHODS

There are a number of processes and associated methods any organization uses. This brief overview is further developed in the referenced material. For a deep dive, readers are referred to them.

As the title implies, this is the business side of the High Reliability house, i.e., manufacturing, service, customer facing, etc. There six high-level categories that apply to all organization, whether for profit or not.

- **Business Process Management:** The ecosystem of all workflow and/or task processes, internal as well as external and required metrics (KPIs).
- **Marketing:** Classic marketing (including digital) and Brand management. Position the firm as Smart Manufacturing.
- **Economic Development:** Impact the firm makes on local communities including HR sourcing as well as materials and third-party services.
- **Training:** State-of-the-Art training, i.e., instructor led, eLearning, etc. to attract, reskill, upskill, and retain a workforce possessing the necessary KSAs.

- **Sales:** Successful sales processes are changing, and firms must have successful selling processes in place to generate revenue, especially from newly developed products and service.
- **Safety Culture:** A robust Safety Culture with appropriate processes such as a Safety Management System is imperative.

As noted earlier, HRM thought leadership is imperative for Smart Manufacturing firms and those that support or provide goods and services to the manufacturer. Expect this requirement to grow as industry rolls out this transformation.

2.8.2 ENABLING TOOLS

Supporting the business are sets of existing and very new and exciting technologies that are necessary for attaining and sustaining Smart Manufacturing. Some key ones are provided herein and as always, the list is not all inclusive.

- **Serious Games:** Online gaming are primarily training but nature, but some are emerging the help organizations address regulatory and social issues. One example is the Cross-Cultural Online Serious Game design for team and collaboration training but recently adapted to help management meet Diversity, Equity and Inclusion (DEI) organizational mandates.[53]
- **Operations Management System (OMS):** For example, according to Chevron, "OEMS (Operational Excellence Management System) is a management system designed to establish and sustain preventive and mitigative safeguards and to assure that these safeguards manage risks and achieve OE objectives. There are two critical aspects of managing safeguards: establishing safeguards and sustaining and assuring safeguards."[54] Moreover, the OMS must integrate the SEMS into seamless workflow, regulatory and reporting processes.
- **Process Simulation:** The use of simulation models and similar tools can add value throughout the manufacturing facility lifecycle. Much like airlines train pilots on most possible scenarios, these tools are finding use in industry.
- **Business Analytics:** This category includes any and all software and other solutions designed to analyze and/control all aspects of Operational Excellence processes, i.e., AI, VR, Augmented Reality (AR), even spreadsheets, etc. Cyber security might be included here or as part of Strong Bond Governance.
- **Assessment of Processes:** Candid ongoing assessment and review of existing processes with an 'eye' toward continuous improvement as well as upgraded or changed based on production needs et al. must also include the economics. Business, technology, and economics are always part of the mix, and this basket of metrics will help drive more robust and safer practices.

Smart Manufacturing can be viewed as a decision support solution. Whether human or digital, assessments and judgments are made as a result of data inputs. One of the challenges faced is the wide variety of data sources and often from different legacies. Most large organizations have a number of systems, often from different eras and IT architectures that must work together.

The above figure depicts a second tier for both Process Management as well as Enabling Tools. These key subprocesses and (some) of their subprocess follow:

It is important to note that a full assessment of an Operational Excellence model will review a significant number or processes, requirements, tools, and data. Within the scope of this book, we are limited on the level of detail. Keep in mind, the transition to Smart Manufacturing is a major initiative and proper planning and risk assessment are paramount.

Missing or understating an issue can have major consequences once operations/production comes online. There is even less room for error going forward.

2.8.3 STRONG BOND GOVERNANCE

This model or organizational governance is akin to the strong atomic bonding in nature as opposed to weaker chemical molecule bonding. As the number of high visibility industrial disasters continued, it became apparent that it was no longer satisfactory to simply empower field management. The risk to the very corporation drove operational governance par with Sarbanes-Oxley (SOX) Act of 2002 following multiple financial collapses the fraud at the beginning of this century.

The model was first published in 2014 in response to issues raised by the Deepwater Horizon incident in 2010.[55] Concurrently, Asset Integrity Management was emerging as a major issue AND another weak link in a governance chain for

TABLE 2.1
Key OE Processes and Enabling Tools

Business Processes and Methods	Enabling Tools
Business process management	Serious games
• *Standards*	• *Cross-cultural engagement*
• *Regulatory*	• *Sales*
• *Metrics*	• *Operations*
• *Ecosystem*	
Marketing	Operations management system
	• *Smart Op Ex*
	• *New build*
	• *Asset management*
	• *Disaster recovery*
	• *Inspection*
Economic development	Process simulation
	• *Workflow*
Training	Business analytics
• *Instructor lead*	
• *eLearning*	
Sales consulting	Assessment of processes
	• *EVPM*
Safety culture	
• *Change management*	

many industrial firms. As such some organizations codified asset integrity into governance models.[56]

Recently, the new EGS model builds on these constructs. Senior management must be directly involved and cannot delegate decision-making when the risks are so high.

When the result of failure can be cataclysmic or at least result in significant economic loss, an organization has the ability for a 'warlike' negative impact. These stakes are high and will get higher—truly bet your company and may be even criminal charges for individuals.

2.8.4 DATA LESSONS FROM COVID-19

Big Data and a host of applications draw on data lakes and warehouses. The Lakehouse Architecture, a new open model, combines the best of both.[57] In the Risk Mitigation section, we will discuss data issues in greater detail.

Moreover, massive data are acquired from a variety of sources, tools, historical data sets, and the list goes on. Many pundits, consultants, and data providers discuss data issues as if they are solved and not that big a deal. Yet, fortunes are spend managing data. We have already noted significant failures based on data analysis.

Smart Manufacturing success will be largely dependent on the timeliness and quality of data (real time and stored). This appears to be an open and unsolved problem. All global sectors have direct, immediate, and very visible access to a major data management problem with massive cost ramifications both in national wealth and human lives—Covid 19.

Everyone knows that since the beginning of 2020, world populations have been living with a dangerous coronavirus. Historians and academics will publish a great deal on the pathogen and the response to it.

However, one major immediate take away that needs to be addressed now is the collection and management to *big data* associated with the disease and its management. Similar issues must be addressed for any meaningful climate change strategies.[58]

For both challenges, data were collected and managed from a plethora of sources with apparent different standards. Moreover, likely bias was introduced in some cases to serve social and political agendas. In addition to the detriment of decision-making, many lost faith in the advice and subsequent regulatory requirements issued by many different levels of government and associated agencies.

The data issue will be further addressed herein. However, it is not clear if the necessary data are at a maturity level that mitigates risk associated with. There is much more work to be done on this subject.

2.8.5 FINALLY

Successful Smart Manufacturing–based firms will have met the challenges of disparate, often legacy-based process integration and optimization. Moreover, reliable and valid data are critical to successful and safe decision systems.

For some firms, this component of the Roadmap may be the most challenging. Business and technical processes coupled with the appropriate enabling tools and quality data are the backbone. This is not the area in which to take shortcuts.

2.9 STANDARDS

Standardization is an important part of this digital-driven era. Moreover, they enable economies of scale along with repeatable processes that drive down cost and drives up value. Simple example: The use of 120 v, 60 Hz electricity (with type A and B plug types) used throughout North America enable consumers to purchase electrical equipment from a wide number of suppliers.

However, standards are not static; they exist at stages in the lifecycle of current socio-technical situations. In some cases, they are frequently updated as the technology and Use Cases evolve. As with science, there are 'no settled standards.'[59] New innovations required industry sectors and other standards bodies to review and update guidance or requirements in light of the new understanding.[60]

Note: At the end of this chapter, there are three tables relevant to this issue; 'Industry Organizations and Standards Bodies and Research,' 'Advisory and Certification Services,' and 'Relevant Standards.' While not all inclusive, these lists can guide researchers and practitioners toward relevant requirements to implement and manage a successful Smart Manufacturing firm. Standards play a significant role in the Smart Manufacturing Roadmap.

2.9.1 MANUFACTURING

Standards tend to emerge after changes have been made either in technology or business practice or both.[61] The manufacturing sector has a long history of innovation. In more modern times, some of the processes still used can be traced to the early 20th century and the advent of Scientific Management.

None other than the esteemed management consultant, Peter Drucker has said that Frederick Taylor was the first to undertake a serious study of 'work.' Once the construct of work was codified, standardized processes, protocols, and regulations could be developed and applied across sectors. The process has continued since Taylor published, The Principles of Scientific Management, in 1909.[62]

2.9.2 INFORMATION TECHNOLOGY

There are a large number of IT standards such as the Extensible Markup Language (XML) and ISO/IEC.[63] They are numerous and in most cases IT vendors are knowledgeable and have incorporated relevant specifications into their product and solutions offerings.

Moreover, major IT providers have de facto standards, i.e., Microsoft Windows, Apple IOS, Google, etc. During the technology assessment process, these guidelines/requirements will need to be taken into consideration.

This is a major IT subject in and of itself and is outside the scope of this book. Practitioners are advised to assure that this issue is addressed directly in their Smart Manufacturing Reference Architecture and Roadmap.

2.9.3 STANDARDS MAPPING INITIATIVES

In 2016, the US National Institute of Standards and Technology (NIST) acknowledged that, "Smart manufacturing, different from other technology-based manufacturing paradigms, defines a vision of next-generation manufacturing with enhanced capabilities." They recognized that existing standards and requirements as well as new ones would need to 'work' in this new environment. They published a mapping of standards across manufacturing lifecycle dimensions; product, production systems, and business strategy and processes.[64]

Other similar initiatives such as reflected in ISO/IEC TR 63306-1:2020 Smart manufacturing standards map (SM2). This effort focuses strictly on Smart initiatives.[65]

Moreover, there are efforts underway to 'harmonize' standards across the globe. The intent is to minimize duplications and inconsistencies. This should help simplify the compliance issue.

2.9.4 VOCABULARY

For over 11 years, the firm MTConnect has provided extensible domain-specific vocabulary and data model that combined with other standards is used on more than 50,000 devices in over 50 countries as well as in thousands of software solutions by over 300 suppliers of IT solutions providers.[66] There are many other entities, industries, and others seeking to develop and codify a common vocabulary.

For example, the Pilot to Air Traffic Controller needs a common understanding of critical terms. To that effect and in conjunction with other entities, the US Federal Aviation Administration (FAA) has published a glossary.[67]

This is a very important activity, especially with rapidly emerging technologies often using very similar acronyms and so-called 'buzz words.' In addition, with heavy industry critical processes such as manufacturing a Safety Culture agreed upon lexicon is necessary as well and this subject will be further developed herein as well.

2.9.5 CONCLUSION

The role of standards is becoming more important; however, like many things attempting to regulate technology the often lag and do not represent the current situation. Often conflicting they are an important consideration.

For example, the Right (License) to Operate is not just a legal term, it is the lifeblood of the firm as well as a component of ESG. Without it, there is no business for the facility/organization in question.

Smart Manufacturing is at the nexus of business, OT, IT, and social written standards, regulatory requirements, and local perceptions. It is imperative that businesses be well run and in accordance with applicable standards and Generally Accepted Practices (GAP).

2.10 SUPPLY CHAIN MANAGEMENT

The Covid-19 pandemic event beginning circa early 2020 is revealing the importance of global supply chains. Perhaps, more importantly, weaknesses and gaps became exposed and negatively impacting both businesses and end-user consumers.

Shipping delays, labor constraints, parts availability, and host of other problems caused by the pandemic shutdowns led to significant disruptions that lasted months. The lessons learned from the events of 2020–2022 will keep academics and consultants busy for years.

For those involved in multiple site Smart Manufacturing, this issue remains paramount for operational and bottom-line performance and overall strategic success. One can argue that SCM is core to successful Smart Manufacturing processes. In this section, we will develop this part of the Roadmap.

Moreover, disruptions at the magnitude discussed above would most likely eliminate any value derived from Smart Manufacturing initiatives. Risk mitigation models will need to be incorporated into all operations, as is the case with other critical organizational processes.

2.10.1 ENTERPRISE RESOURCE PLANNING

Most organizations have implemented ERP systems to enable the daily operation of the firm and even its supply chain. Many of these systems are well established having been deployed circa 1980–1990s, although new ERP systems are still being deployed.

Many argue that older systems should be replaced as new technology becomes available. Often, this is not feasible. For example:

- Systems may not have yet been depreciated and software fully amortized.
- The organization may have a lot of experience with these systems and the costs of retraining large numbers of users may be prohibitive in time and capital.
- Data formatting and compatibility with older systems can be challenging in terms of time and financial costs.
- Lastly, political factions may need to be addressed.

Smart Manufacturing deployments will need to be integrated with these legacy systems as required to assure a high enterprise-level performance. The integration issue is one of the factors that must be considered in risk models.

2.10.2 USER INTERFACE

As we addressed in the Human Factors section, it is that important human beings be able to interact, interpret, and respond to data-driven requirements. To avoid the list of passwords on the yellow sticky pad pasted on the monitor, the IT sector developed and continues to enhance methods to simplify complex processes.

One we all can relate to, the concept of a Single Sign On (SSO), has been around for a couple of decades. This allows users to log into a number of diverse, discrete software products using one profile and password. While there are strengths and weaknesses to this security model, the idea allows users to access relevant data and information easily.

This approach is popular and while it might be automated, the human sees this as a productivity tool and when the applications are seen as an integrated suite, it makes perfect sense even if the software is from a number of independent vendors.

Similarly, software vendors have long pushed integrated solutions that enable data sharing across their product lines and solutions. Microsoft Office is one example that most are familiar with. Logging into MS Office provides users (based on licensing) access to a plethora of software products. Most major providers offer subscriptions to their integrated software suits.

Colloquially, this is the concept of 'One Throat to Choke.' In other words, the customer has only one vendor to look to for support, troubleshooting, etc. This concept may no longer be as relevant, and its demise may create another risk exposure.

Smart Manufacturing architectures most likely use products from a number of vendors and in some cases use Open-Source products. Multiple vendors can lead to issues when the system does not perform as anticipated or failures are difficult to troubleshoot, i.e., finger pointing.

2.10.3 SCADA

In the plant environment, Supervisory control and data acquisition (SCADA) has been core to process manufacturing for decades. As such, there are several if not many generations of technologies in place.

Manufacturing firms are risk-averse for a number of valid reasons. There is risk associated with change and this must be balanced with the benefit of change.

As such, generations of SCADA technologies will need to be integrated into Smart Manufacturing. There may be 'greenfield' or clean slate opportunity when a plant turnaround replaces/upgrades many decades of production facilities. These systems must be integrated into new business and technology models, the approach is addressed herein.

2.10.4 DIGITAL THREAD

This concept is a communication framework that facilitates data transfer across silos and throughout the asset lifecycle. Its intent is to make data available on demand to those that need it in a seamless manner.[68] Data at the 'Right Place at the Right Time' remains a significant challenge, and in many cases the data may not reflect reality—valid and reliability. This issue is decades old and shows no sign of resolution.[69] Beware is the watch word.

2.10.5 DIGITAL TWIN

The digital twin is a virtual model of a physical asset. Sensors input real-time data with the goal adding value to the physical item.[70] This approach allows for cost-effective and rapid changes in engineering revisions and upgrades to product development as well as manufacturing processes based on the said equipment. In other words, digital testing of the new without significant investment in physical prototypes.

2.10.6 BLOCKCHAIN

As supply chains extend globally with growing numbers of contractors, subcontractors, suppliers, and even small vendors participating, financial transparency becomes more important. Each 'time stamped' transaction block is linked using cryptography (secure communications) and is resistant to modifications. Any changes to one block are reflected in all subsequent blocks.[71]

For manufacturers of all sizes, blockchain enables trust and workflow automation. This technology helps strengthen trading relationships, tracking components, and final goods throughout the "chain of custody."[72]

This provides greater security against fraud as well as an audit chain. It appears to be the technology that is being widely adopted.

One final thought, industry standards are emerging. For example, the Global Blockchain Business Council claims the map data from almost 400 groups.[73]

2.10.7 CONCLUSION

Given current events, the pandemic, geopolitical concerns, labor and material shortages, we can expect the dynamics of supply chains to continue to remain challenging. The key role this process plays with Industry 4.0 firms requires robust agility, resiliency, and sustainability.

In the remainder of this chapter, we will develop the tools and solutions that will help organizations develop a high reliability of supply chain performance. The Capstone Roadmap will provide guidance toward success.

2.11   RISK MITIGATION

Exposures and subsequent risk mitigation practices can be very complex for Smart Manufacturing–based organizations. Yet, like many concepts, it is difficult for humans to conceptualize all the dimensions of hazards.

There is a tendency to simplify risk models and the following matrix is one popular example. As noted earlier, the concept of High, Medium, Low or Unlikely, Likely, Very Likely as a model is ingrained. Often as a result of management's desire for a 'high level' presentation, details are overlooked or even hidden. This can lead to a level of false security.

Things are rarely simple, especially with complex systems. There is a large body of empirical work that documents causation of major industrial incidents and in most cases, there is not only a single cause. Rather, it is a combination of issues and reactions to concerns that lead to a major incident. Some are detailed further in this chapter.[74]

For example, even a simple matrix such as the one above generates a number of possible outcomes. "The fundamental source of that complexity was a phenomenon well known to systems engineers: the number of potential pairwise interactions among a set of N elements grows as N times N-1, divided by two. That means that if there are two elements in the set, there is one potential interaction; if there are five elements, there are ten potential interactions; ten elements and there are forty-five; and so forth. If the interactions are more complex, such as when more than two things combine, the number is larger."[75]

Smart Manufacturing risk mitigation practices require more robust and dynamic, aggressive risk management. This section will detail some key issues and approaches.

However, keep in mind that industrial, enterprise risk management is a major subject and beyond the general scope of this book. Readers should take full advantage of the knowledge base and not rely exclusively on this high-level review of noninclusive list.

Per the PMI, all projects and organizational transformation initiatives involve risk. Risk can take on many guises, namely, business, technical, social, etc. These

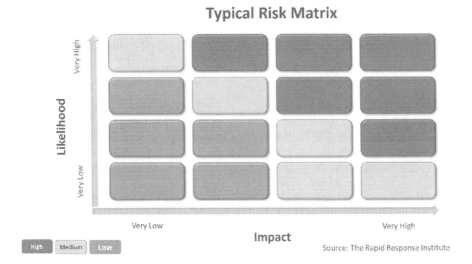

FIGURE 2.12   Typical risk matrix.

potential exposures require that a mitigation plan be put in place and implemented as required.

This requirement is part of the HRM model as previously mentioned. Perhaps, the largest systemic and ongoing risk a Smart organization faces revolves around data acquisition, assessment, and decision-making based on the validity and reliability of data input and any associated uncertainty.

2.11.1   BARRIER RISK MODELS

Preventing an incident is the goal of all and responding effectively when sometimes does go wrong is consistent with HRM. Two other popular models include the Bowtie and its partner Swiss Cheese Barrier constructs.

- **Bowtie:** Common to engineering designed and used by some as a daily operational tool, it consists of three components. The hazard or potential incident Event and the Threat/Cause as well as the Effect of barrier failure and Response. One caveat: In the author's experience, some firms only use this model during the design phase. If it is used as a daily operations tool, it must be continuously updated. Several firms and software providers have expertise and appropriate tools available.
- **Swiss Cheese:** James Reason enhanced the barrier paradigm arguing that any barrier can have weaknesses or 'holes' and if multiple barriers (defense in depth) simultaneously fail (the holes line up) an Event will happen. The goal is to eliminate this potential alignment.[76]

Rapidly responding to barrier(s) failure is a solid idea. One trial to designers; have we identified and properly engineered all the barriers. We will discuss latent variable shortly—the message ALL barriers can never be knowable. A broader approach to risk management is required.

2.11.2 ENTERPRISE RISK MANAGEMENT

Since Smart Manufacturing will be the 'way we run our business,' it is logical that the mitigation of resulting risks must be not only aligned but integrated into the Enterprise Risk Management (ERM) platform.

As shown in the following figure, the Risk Governance Model with a focus on operations is the most robust approach. Following the ERM model put forth by the Committee of Sponsoring Organizations of the Treadway Commission (COSO), this construct was developed about a decade ago as the author and his colleagues were helping clients integrate SEMS into existing OMS.

While this issue will be further developed, suffice to say that governance at the time was more focused on finance coming off so the scandals at the time, i.e., Enron, MCI, et al. identified operational weakness were identified and early asset integrity integration models developed.[77]

The components in the model are described in depth throughout this chapter so they will not be further addressed herein. However, note that this framework is detailed and addresses all known issues the organization is facing. This makes it a good one for the transformation to Smart Manufacturing.

This model is but one approach. Others are equally of value and for any given firm another choice might be better.

Most large manufacturers will have already implemented an ERM solution in place. All major software providers and many others offer solutions. The main point—operational risk management must be aligned and treated as it is integral to ERM. Because it is 'the way we run the business' going forward.

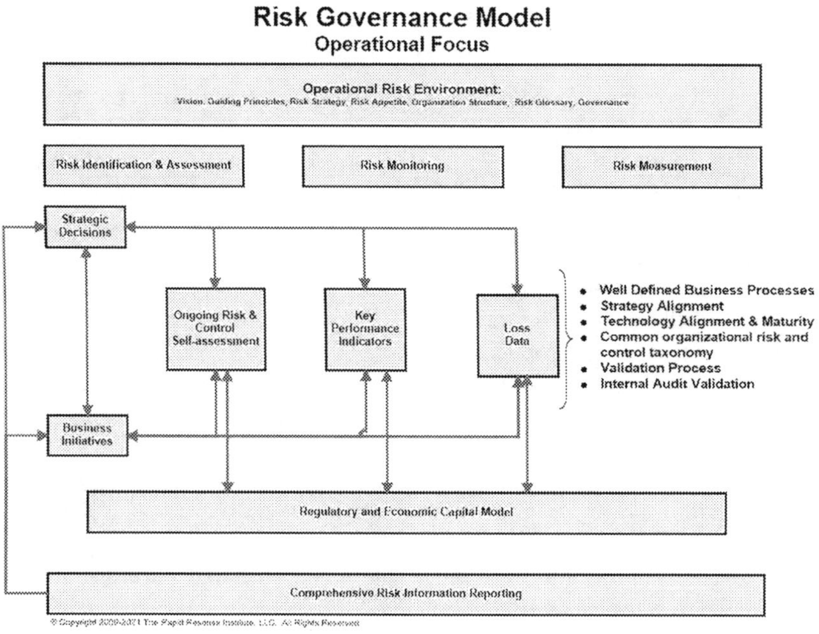

FIGURE 2.13   Risk governance.

2.11.3   SYSTEMIC ERRORS OR DATA BIAS

All decision-making is based on incomplete and often conflicting data inputs. It is the human judgment upon which success or failure rides.[78] This judgment can be clouded by data bias.

The fundamental question that must be answered for all data; is it Reliable and Valid? If not the adage, 'garbage in—garbage out' applies and any results or assessments are questionable.

Data errors and errors are well known and, in some cases, have led to major failures or even to the brink of Armageddon. In 1983, an issue with the Soviet Union's early warning system indicated missile launches by the West. Fortunately, the duty officer was knowledgeable enough to believe what he was seeing was suspect and overrode the system. Turned out satellite sensors were detecting sunlight reflecting of cloud tops. A widely not known close call.[79]

Issue regarding data and measurement integrity cannot be overstated. Particularly as organization run their company using Smart technologies and processes. Moreover, there is US Security and Exchange Commission (SEC) guidance precedent for public companies, municipalities, and others regarding exposure to IT systems. In 1998, the SEC issued 'guidance' effectively required organizations under its authority to disclose any 'material' issues and exposures to Year 2000 Issues and Consequences.[80]

It is reasonable to expect that as organizations depend almost exclusively on Smart to run the business, additional guidance and regulations will emerge. This underlines the seriousness of data management today.

Typical data bias includes:

- **Information:** Distortion in measurement possibly due to poor instrumentation whether psychological and senor.[81]
- **Selection:** Randomness or a nonrepresentative sample is not attained.[82]
- **Confounding:** Distortion of a data relationship where none may exist.[83]

Additional subgroups of data bias have been identified but all appear to be components of these high-level types.

Most have heard of the Covid-19 False Positive and False Negative testing. This is a good example of Type I and Type II errors. In this case, a medical sensor is in error. We have seen the resulting impact on individuals, organizations they belong to, and the general public as well as policy decisions.

Sensors and other Smart Manufacturing data sources can be error prone. The result may be either type error as feed into decision-making systems.

In the event of a catastrophic error(s), systems can become overwhelmed and either make bad decisions or humans with choices that are not well understood. When multiple inputs fail or are compromised, systems can rapidly get out of control and go into a nonrecoverable state.

Industry safety is replete with case studies of massive failures of this nature. Cascading failure can occur in any interconnected system such as the electric power grid, transportation, and financial systems.[84]

Smart Manufacturing has exposure to these and other risks. Mitigation efforts must take these issues into consideration.

2.11.4 LATENT VARIABLE ANALYSIS

One other important bias that must be considered by data scientists and others is what is known as *Survival Bias*. Latent variables are those that are inferred from measured data. This statistical model came about during World War II when the statistician, Abraham Wald noticed that bomber aircraft returning had noncrippling damage. He surmised that those aircraft that did not return were damaged in other areas. This analysis resulting in changes being made to aircraft armament.[85]

Latent variables are critical in Structural Equation Modeling, a method for assessing relationships among behavioral variables. As hard and soft data coalesce in the IT-OT model, knowledge of their interaction and output will be of increasing importance in Smart Manufacturing behavioral systems.

Finally, one should assume that ALL data have bias or are incomplete or even wrong. Effective mitigation strategies by data scientists and executives are mandatory.

Smart Systems, "Won't privilege certain types of data or results out of personal loyalty. It won't refuse to see a pattern because it's emotionally attached to a different point of view. It will, however, create rules based on whatever data you feed into it. If that data tells a skewed or incomplete story, the rules it creates will be based on these foundational errors."[86]

This last statement/quotation is *very important*. In other words, 'agendas' can find their way into systems and distort the output to the narrative of interest. Passing the so-called, 'smell test' is a human override.

2.11.5 CYBER SECURITY

It goes without saying that cyber security is paramount in Smart Manufacturing. While 'bullet proof' may not be attainable at present, best-in-class capability is necessary when digitalization is 'how we run the company.'

Fortunately, there is a growing body of knowledge enabling organizations to mitigate this risk and rapidly respond if the worst happens. While the authors do not advocate the following vender and its solution, this is a good case study of available cyber risk mitigation tools and processes.

MITRE Corporation is a working, "at the intersection of advanced technology and vital global concerns since our founding in 1958 as a private, not-for-profit company providing engineering and technical guidance for the federal government."[87] The firm has developed a knowledge-based cyber taxonomy, ATT&CK®, which documents tactics and techniques cyber adversaries may take.

The authors do not recommend any specific product or service described herein but interested parties may want to explore this service more thoroughly. Of relevance to a Smart Manufacturer includes ATT&CK for:

- **Enterprise Matrix:** Attacks at the enterprise level.[88]
- **Mobile Matrices:** Android, iOS devices.[89]
- **Industrial Control Systems (ICS):** Knowledge base for most operating inside an ICS network.[90]

There is a large body of knowledge of which MITRE represents one source. The challenges for security of an online enterprise are great, evolving rapidly and can cripple an organization, i.e., the 2021 ransomware attack on the Colonial Pipeline. Interestingly, that adversary used public knowledge provided by an antivirus firm to fix a flaw in their code and then launch the 'successful' attack.[91]

Included in the emerging body of knowledge are a number of standards developed by a variety of organizations. Interested parties should research this knowledge base for those that are relevant to their organization.[92]

2.11.6 PERSONNEL AND EQUIPMENT SHORTAGES

The Covid-19 has stressed the global supply chain, most visibly the 'Computer Chip Shortage' which by some accounts may impact on the availability of smart devices, i.e., sensors, controllers, analytics, etc.[93] Delays in required hardware may impact transformation schedules and perhaps the nature and scope of the original Proof of Concept (PoC). Hopefully, as the full rollout begins, much of this supply chain issue will be mitigated and normalcy will return.

Throughout this chapter, the skills required (KSAs) for the future are discussed. Recognizing any gaps that exist and processes to address these holes are also part of the risk mitigation profile.

Most likely, the human resources issues will persist. Despite STEM and other initiatives, most likely HR will be the key limiting factor in this transformation and sustainment of Smart Manufacturing.

2.11.7 INTEGRATION WITH THE EXISTING SYSTEMS

Like people, IT systems have history. Organizations have been using electronic IT solutions to manage business processes for at least six decades. Early systems used punch cards to hold data.[94]

Legacy systems are part of the landscape. In some cases, significant capital investments were made, and organizations built their business around 'at the time' state-of-the-art solutions. Personnel are trained and comfortable with these systems and for the most part they work reasonably well.

Some talk of a mythical 'greenfield' whereby Smart solutions can be deployed in a vacuum. The reality is that new solutions using new technologies must co-exist with the past.

One key issue with IT is the advancement of software, its coding language, and architectures. Often referred to as legacy systems, they are the present and, in many cases, have not been fully depreciated (hardware) or amortized (software). Therefore, it is not easy to retire them financially with significant retraining costs.

Enterprise-wide systems are a combination of multiple generations of technology often referred to in 'dog years' due to the rapid development of the new. Moreover, multiple Smart Manufacturing software vendors are now using Open-Source software to develop their product lines. This may create intellectual property (IP) ownership issues or perhaps more damaging cyber security risks.

As with most risk issues in this transformation, this is a big one. Significant effort must be expended to develop and implement this component of risk mitigation.

2.11.8 PERCEPTION

Long time slang for the acronym CIO is 'Career Is Over.' This speaks of challenges that IT projects have come to represent over several decades. When the early ERP systems were developed and deployed, it was necessary to have individuals who understood business processes assigned to the project. This was meant to assure that the end state system actually addressed a 'real' problem.

On the surface, this made sense; however, from this author's perspective many businesspeople did not believe that working on the project was adding value to their careers. In fact, many felt it negatively impacted on their opportunities for advancement. Major cost overruns did not help this view.

Rather than having P&L responsibility, operations personnel were sidelined into projects with known challenges. This reputation continues to this day.

It is important that management convince 'line' personnel of the value of their participation in IT-OT based initiatives. It is the integration of operations and IT that assures that careers are enhanced by those teaming together to digitize 21st century firms.

2.11.9 SYSTEMIC VS. SYSTEMATIC

According to Investopedia, Systemic Risk is defined as a broad, major shock to a system. Examples include the financial system collapse of 2008 and more recently, the Covid-19 global pandemic. Systemic exposures of this nature should be part of the ERM Framework.[95]

Systematic Risks include those caused by complex, interconnected, time-dependent processes such as economy, geopolitical and now Smart Organizations.[96]

The Smart organizational construct is relatively new; therefore, not much of a track record exists. However, organizations can learn from the mistakes of others and there is a substantial body of knowledge regarding complex systematic failures with major consequences, i.e., the Titanic, Deepwater Horizon, Three Mile Island, Space Shuttles, and so forth.[97]

The Uptime Institute posits that "The hallmarks of so-called complex systems are a large number of interacting components, emergent properties difficult to anticipate from the knowledge of single components, adaptability to absorb random disruptions, and highly vulnerable to widespread failure under adverse conditions." In addition, the components of complex systems typically interact in a nonlinear fashion, operating in large, interconnected networks.

Moreover, organizational ecosystem behaviors can be contributing factors. For example, a Systems Analysis of the Deepwater Horizon incident of 2010 uncovered several process failures shown in the following list.

Notifications: The business process and enabling systems in place did not properly notify individuals that were either management or those that could add value to the scenario. This was true well before the unfolding events of that tragic day.
Information Sharing: Data and engineering results were not shared between those groups/companies that needed these results.

Workflow Changes: As a result of engineering changes in advance of the incident, changes were made by more than one individual and department. This was not transmitted to those parties/suppliers who had a need to know.

Escalation: Systems were not in place to escalate issues when identified and not acted upon.

It is important to note for this major disaster, the supply chain was heavily involved. For example, in addition to the oil company operator, the drilling contractor as well as a major energy services firm were directly involved in the decision-making and actions. Additionally, subcontractors, suppliers and others were engaged as well.[98]

These Issues are Faced by All Firms working in Critical Infrastructures and Can Be Very Complex! Intelligence machines making autonomous decisions only add to the level of complicated system-wide behaviors.

HRM will help keep the level of exposure lower. Do not underestimate system risks.

2.11.10 AI, MACHINE LEARNING

The October 2011 social media reveal regarding issues with young women dealing with personal issues, frankly, is not new. One of the classic Big Data failures is said to have occurred in April 2011 when the book, *The Making of a Fly* was listed on Amazon.com for prices ranging from \$35.54 to \$23,698,655.93 (plus shipping). Erroneous models were to blame at the time.[99]

Moreover, the author has addressed issues with algorithms a number of times and there are ample documented case studies to support concerns raised.[100]

We can all relate to the 'moving' price of travel. Specially, airline prices seem to get more expensive with each 'search.' It appears that the more times an IP address hits a desired destination the more expensive it gets. While this may help the vendor maximize profit, this perception (risk) causes some customer angst if not a greater reaction.

In October 2021, a social media whistleblower raised issues regarding major outlets and their impact on the self-esteem of teenage girls. Specifically, the role of 'machine-learning algorithms' used by social media. Machine learning algorithms are trained, hence their ability to automate data served to individual users and their demographics. The expectation is predictive in that if a user likes to workout, he/she might be interested in gym equipment or other exercise programs.

Some points are well known and by using the sites users accept and even welcome their outputs. Others are not as public. Examples include:

- User (presumed) *Preferences* based on past performance, i.e., products, travel etc.
- *Specific Bad Content* including spam, inappropriate behaviors, etc.
- *News Feeds*, i.e., type of stories of interest.
- *Increase Engagement*, i.e., number of likes, shares, etc.

As we have seen from politics, increased engagement can cause polarization in societies. Moreover, "The machine-learning models that maximize engagement also favor controversy, misinformation, and extremism: put simply, people just like outrageous stuff."

Before, the US Senate, the whistleblower indicated the engagement base ranking system is negatively impacting young women. Those who have a tendency toward melancholy content are at the greatest risk.[101]

Increasingly, libraries of machine-learning algorithms are being adapted to new applications by engineers working on specific processes. Given the critical nature of many of these data feed software applications, oversight or Quality Assurance processes need to be in place to assure calculations and decision support solutions are correct.

Finally, all parties involved in the development process will need to be part of an ongoing enterprise-wide training program in this field. This is especially true for new engineers who may not have the depth of organizational engineering knowledge.

What do you do if your machine is learning the wrong things?

2.11.11   GLITCHES

Recently, as a result of *flickers* in electricity, this pundit's Internet (including Wi-Fi) went down for almost 30 minutes—the rest of the power to the facility remained online. An inconvenience, yet irritating. What if this *sensitivity* happened to a Smart Manufacturer? The results might be more than an annoyance.

Cloud systems have low tolerance for voltage and amperage 'flickers.' Something to consider in your risk mitigation strategy. Thirty minutes can be a lifetime in a production environment. Consider uninterruptable power (UPS) as a minimum.

2.11.12   FINAL THOUGHTS

Risk is a broad and very important subject, and its effective management is critical as firms transform to Cloud-based business models. Cyber breaches and major industrial incidents are well known and the focus of media and shareholder interest.

Daily risk 'blocking and tackling' is the nonglamorous tasks that help prevent and mitigate the later. One of the most important aspects of Smart Manufacturing if not the most important. Interested readers may want to review, ISO 31000:2018 Risk management—Guidelines for further insight.

2.12   ENABLING TOOLS

The conversion to Smart Manufacturing is by definition a *digitalization* process. From one perspective there are two major tool sets necessary for a successful transformation: Software and associated processes as well as Value Assessment.

This section is a high-level overview of various tools necessary and is meant to provide a set of guidelines for use in the Roadmap development process. It is not meant to be all inclusive but a companion to other chapters in this book.

2.12.1   SOFTWARE

There is a long and robust number of key software tools and solutions that add significant value to Smart Operations. Interested readers are invited to do an online search for current details.

In this section, we address the software Technology Assessment process. In other words, which solutions add value and how they fit into your existing platform.

The following figure depicts the relationship between the Process Maturity previously discussed as a function of the software Technology Maturity. Moreover, no software deployment exists in a vacuum as they must fit into an existing platform.

Regardless of the quality of the governance model, software implementations are built around these models including ESG. Likewise, change management is always part of the ongoing process of assimilating new technologies. At the highest level, this integration adds to organizational and even sector Knowledge.

There are two axes, the level of process maturity as function of the levels previously identified. Technology also has a maturity growth. The length of the vertical oval is meant to show which step of the process maturity the technology has a good fit with. Some like cyber security are critical at every level of maturity. Others fit better with more mature organizations.

Moreover, from left to right is the function of maturity of software. For example, Statistical Reliability Analysis contained some of the newer solutions such as AI, and Big Data, where the maturity of the code is younger than more established solutions.

This is not meant to suggest that a less mature code is less valuable, in fact in some ways it may be of greater value. It is simply meant to show that some software is earlier in its lifecycle than others.

However, newer models still can have issues as they move from adolescence to a higher level of maturity. For example, Stanford recently announced their 'Foundation Model' for AI. This announcement was quickly met with a negative response from equally credentials individuals.[102]

It is beyond the scope of this book to address the details of this disagreement regarding new technology. It is simply to document those newer technologies continue to go through a vetting process until they have been used extensively.

We briefly define these key technology components of Smart Manufacturing as the following. This list covers the major categories of technologies but may not be all inclusive. It is meant to illustrate the suite of technologies necessary to manage the modern manufacturing-based corporation.

- **Cyber/Physical Security:** Critical to all online or Cloud-based information systems, security of data, information, and decisions taken therein must be protected. However, given the state of hackers so well documented recently, this technology is relatively immature yet and should develop rapidly.
- **Statistical Reliability Analysis:** Reliability is the linchpin of any HRO. Moreover, stochastic assessment is the preferred way to analyze any given situation. Tools such as AI, IoT, et al. fit within this construct.
- **Integrated Ops:** For decades, management has recognized the value from integrated operations and the ability to seamlessly share data among applications and individuals/teams. SSO is an enabling tool.
- **Sys Portfolio Mgt:** Most large organizations view the product/service lines from the perspective of portfolio management. Typically, the concern must be large enough to perceive its portfolio from a systems standpoint.

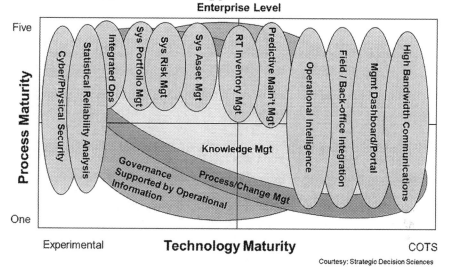

FIGURE 2.14 Key technology/process profiles.

- **Sys Risk Mgt:** As previously discussed, systemic risk management is paramount in modern global businesses as well as government agencies.
- **Sys Asset Mgt:** A firm is a set of revenue-producing assets, i.e., manufacturing facilities. While each asset has a set of performance criteria the top and bottom lines are functions of the set of the systemic execution of the whole.
- **RT Inventory Mgt:** Recent post-pandemic problems with global supply chains confirm the need for responsive real-time knowledge of critical inventory components.
- **Predictive Main Mgt:** The trend toward maintenance when 'required' is effective with significant cost savings and less unplanned downtime.
- **Operational Intelligence:** Real-time knowledge of the state of operations is critical for all manufacturing firms whether they have one facility or several hundred.
- **Field/Back Office Integration:** A major function within the IT-OT model is the integration of field or manufacturing operations with SG&A functions. The goal is increased efficiency and a better return on investment.
- **Management Dashboard/Portal:** Most organizations provide dashboard access to operational/manufacturing data and information specifically designed for all levels and some include the supply chain ecosystem and even customers. This approach adds to efficiency and better decision-making for all.
- **High Bandwidth Communications:** Access is expected/mandatory, and data and information requirements continue to grow, i.e., Edge computing. 5G technology is being deployed and it is logical that 6G will follow.

As mentioned, this list of major categories is not all inclusive and will continue. As part of the Roadmap section, we will discuss the technology assessment and section process. Moreover, not all software developers are equal, and some may have products that are a better fit for one firm and others for another.

One final thought, some argue that MVP or Minimum Viable Product is an acceptable level for use at the enterprise level. Caveat Emptor on this approach when production and safety are at risk. An MVP is only appropriate for those early adopter firms who have the in-house technical expertise to manage immature solutions.

2.12.1.1 Workflow Engines

Many processes are repeatable and only data inputs change. An easy example is the Expense Account. This is a standardized portal that all employees use and usually provides a printed output, PDF, or simple acceptance notification. Most organizations have many such processes that lend themselves to software.

Earlier we discussed the OMS. This is a good example of a sophisticated enterprise-level process using an underlying workflow engine. As noted, as processes become more complex with significant regulatory oversight, workflow processes assure greater adherence to standards and requirements. They also allow for management oversight, even remotely.

2.12.1.2 Drones

The use of drones in operations is growing rapidly. While this may not be applicable for some locations, those manufacturing sites that are widespread, i.e., refineries, petrochemical plants et al. may find them of use for inspections, security, etc. Data these instruments capture will need to be incorporated into smart solutions as appropriate.

2.12.1.3 Remote Automation and Data Acquisition

There is an old saying in the offshore energy business. Uncited, words to the effect, "the goal is to reduce the workforce on offshore facilities to one man and a dog. The man is there to feed the dog and the dog is there to bite the man if he tries to touch anything." Just short of a fully automated facility.

For many years, offshore operators and service provides have been reducing the number of individuals offshore. The purpose is twofold, first safety is enhanced, and secondly costs are reduced. Moreover, automated systems can increase effectiveness and increase productivity for all heavy industry facilities in the following ways[103]:

- **Increased Precision and Accuracy:** Expect a higher rate of accurate calculations with the resulting better reliability and less human error.
- **Greater Flexibility:** Easily changed workflow with associated business rules, i.e., new product production or enhanced document and data management. Easier goal setting to meet new competitive or regulatory issues
- **Faster Task Processing:** In some cases, humans must still physically go to a site to confirm issues such as liquid or gaseous leaks or confirm vandalism. Remote control and/or real time access to data can positively impact decision-making.

- **Enhanced Collaboration:** During the Covid 19 pandemic, video conferencing quickly became the 'go to' process for in some cases literally keeping business functioning. This and other collaborative processes can expect to be enhanced and expanded.
- **Better Customer Service:** Not only can product configuration and/or customization be enhanced thus assuring better customer satisfaction, but it may provide a competitive advantage. In one recent project, quality assurance of the final rebuild service was tested with customers on video. Previously, clients had physically visited the facility from all parts of the United States. This not only saved the customer time and travel expense, but it also enabled client acceptance and billing much faster as well as freeing up the plant for the next project. Project process time and actual cash flow was positively impacted.
- **Greater Transparency:** As noted previously, there is a major move toward better governance (ESG) and more transparency. Real-time data feeds into advanced AI, AR and Machine Learning tools that enhance visibility for all who collaborate on a given project or process as well.[104]

For those readers interested in additional information about automation, the International Society of Automation (ISA) has published, *A Guide to the Automation Body of Knowledge, Third Edition.* It not only describes important constructs and processes but also provides an insight into the KSAs necessary to accomplish automation tasks.[105]

2.12.1.4   Customer Relationship Management

Some may be surprised that a section on CRM would be included in this book. Many think of this software tools and in the domain of sales and marketing, not manufacturing. Early systems were mostly used by sales and management to manage revenue generation and marketing campaigns.

More sophisticated uses include forecasting production, ordering component parts and other items and processes of manufacturing. Moreover, these systems are now a major part of the digitalization organization.[106]

2.12.1.5   Asset Integrity Management

According to DNV, the independent firm offering assurance and risk management solutions, "Asset integrity management ensures you have the business processes, systems, tools, competence and resources you need to ensure integrity throughout the asset lifecycle. Design, operational, and technical integrity must all be managed effectively to control costs."[107] In other words, this is one of the most critical issues faced by manufacturers.

As a result of a series of major heavily industry serious accidents with loss of life and major damage not just to equipment and facilities but reputation as well, two studies were conducted in 2012 and 2014, respectively. The second was an update of that topical subject and confirmed these three points which are still relevant as of this writing[108]:

- Integrity management is a core part of enterprise risk management, and its role is growing in importance.
- Substantial resources are available to organizations to help them implement.
- Respondents to a survey indicated that in addition to risk management, this is key to HSE as well as production rates and quality.

Previously, we made the case that asset integrity management is so important to the current function and sustainability of the firm, which is the core part of organizational governance. If equipment does not function properly, no amount of software will matter so it is imperative that things like Predictive Maintenance function 'as advertised.'

2.12.2   CONCLUDING THOUGHTS

As stated, this is not an exhaustive list and given the lag between penning this document and its publication, new or additional capabilities may be available. The main purpose is to develop a taxonomy for the deployment of categories of software products, solutions, and suites in Smart Manufacturing.

We will build on this taxonomy and address the risks associated with software deployments in the Roadmap section. The assessment of software technology is a critical part of the Smart Manufacturing implementation and ongoing sustainability.

As a new approach to the business, there are clouds on the Cloud and the haze most likely will not clear anytime soon. Organizations will continue to be challenged deciding on enabling software solutions. Recently, two industry giants put forth the following[109]:

- Use cloud as the foundational technology to enable scale and adoption of built-in data tools.
- Focus on horizontal, cross-plant integration to leverage the benefits of the technology.
- Embrace AI to improve efficiencies and effectiveness.
- Digital transformation prioritization is key to get the biggest return from digital investments.

This list seems reasonable; however, the implied range and depth of software necessary to accomplish this short set of tasks can be staggering. Managing this can be overwhelming. This need not be the case using ideas and concepts put forth in this book.

Technology Assessment is not a one-time event but an ongoing process as part of any Technology Refreshment program to assure software sustainability, i.e., the possible emergence of metaverse—the complete use of *all* available data. After all an enterprise-level investment in Smart Manufacturing may cost more than any other software investment firms have ever made.

While the Board of Directors may not ask about how technology decisions were made, other executives with Profit and Loss (P&L) responsibilities whose divisions will be impacted by this decision may. It is always best to be prepared to address any concerns raised.

2.12.3 Economic Value Capital Assessment

The market research firm IDC expects, "direct digital transformation (DX) investment is still growing at a compound annual growth rate (CAGR) of 15.5% from 2020 to 2023 and is expected to approach $6.8 trillion as companies build on existing strategies and investments, becoming digital-at-scale future enterprises."[110] This is a massive level of investment which creates risks for all stakeholders especially stock and bond holders of firms investing in Smart Technology.

Approximately 20 years ago, following the Enron debacle, studies showed that firms with strong governance outperformed by an average of 8.5%.[111] This is relevant to Smart Manufacturing because it not only ties back to the governance discussion herein. This statement supports the importance of senior management and even Board engagement in operations.

By some accounts, while Smart Manufacturing will dramatically and positively impact production it may in fact negatively impact follow on high margin service revenue streams, i.e., reduction in maintenance profit. Therefore, it is important to take a holistic approach. For example, ways to monetize digitalization investments include:

- Focus on Revenue Growth and not just Cutting Production Costs
- Use Data to Optimize Pricing
- Manage Pricing Scenarios
- Use Data to Segment Customers
- Make the Digitalization Process a Top Management Priority[112]

One aggressive study suggests that the overall value from digitalization at all levels and sectors could be as high as $100 trillion.[113] While optimistic, many believe the return can be high. Yet, how does on organization assess its ROI or ROIC?

It is important to build a Business or Use Case that is credible. One older study suggested that most respondents developed a business case, yet most were not satisfied with the process.[114]

The process itself may be of more value than the final result as it forces the assessment of most variables that may impact on the investment. Keep in mind, this is a time-consuming process when done right.

There are a number of value proposition models available. Most are provided as solutions to address the capital expenditures (CAPEX) for digitalization projects. One that addresses all aspects of the digitalization lifecycle including direct measurable costs as well as intangibles is the EVPM.[115] See the Expected Value of Marginal Information section for additional details.

It is important to take a holistic lifecycle approach to Smart Manufacturing capital investments. After all, it will be the 'way we run the business' for decades going forward.

There must be a bottom-line analysis, like any project. Do not let the hype cloud this judgment and analysis.

2.13   BEST PRACTICES

It can be valuable to learn from others, including from those in different industry segments. However, there can be some challenges when emulating knowledge attained. Each organization is different at both the macro and micro level; therefore, one-to-one comparisons may not be appropriate.

Organizations need to assess how practices, policies, and procedures used by one culture might be valid in their organizational culture. Organizational and sector cultures will be addressed in greater detail in a later section.

Culture can be defined as "who we are." It is a source of pride and competitive advantage. It is what makes the organization unique. There needs to be a clear and well understood path forward when incorporating 'learnings' from others into the firm.

As noted, the construct and Best Practices can be fraught with challenges. Others prefer the term, Good Practices or industry knowledge that may have some applicability to a given problem.

Cautionary Note: Just because a Practice works well for one manufacturing sector or even a competitor in the same sector, does not mean it will work in another without some'tweaking' or more. Transition teams need to be careful and do appropriate due diligence to assure the business model they are adopting is the right one for their particular environment.

2.13.1   OPERATIONS MANAGEMENT SYSTEM

There is a growing need for organization to codify and standardize their approach to operations. In other words, an OMS aka Production and Operations Management (P/OM) System. There are off-the-shelf software solutions available, or firms can develop their own.[116] One Fortune 100 company used MS SharePoint to develop and deploy their OMS.

Manufacturing operations are a very complex process, especially for large global organizations with extensive supply chains. Not only must firms produce and ship goods to customers on a timely basis but also must adhere to various regulations, standards, governance, and safety processes. Moreover, these processes must be consistent from plant to plant and across all suppliers and vendors as well as employees and temporary staff.

Most industries have developed Operation Management Systems (OMS) designed to capture and provide decision support for all aspect operations in a single accessible system. An OMS is a single system that all individuals involved in the process, including the CEO and others, can access the same complex data in the manner in which each individual with his or her role and associated responsibilities can use the output.[117]

Typical elements of an OMS may include:

- **Goals:** Actionable and measurable expectations, including governance policies to assure success
- **Approach:** Organizational Excellence model
- **Roles and Responsibilities:** Assignment to functional, responsible parties including supplier Bridging Documents as appropriate

- **Processes:** Functional workflows, tasks, protocols including feedback
- **EHS:** Required Environmental, Health, and Safety for a true Safety Culture
- **Controls:** Continuous improvement, technical and financial auditing, and compliance management

In addition, OMS must be aligned with the overall corporate strategy and culture including the firms Environmental, Social, Governance (ESG) model.[118] The following figure depicts these components in greater detail and forms the basis for developing, deploying, and/or upgrading an OMS as part of the Roadmap to Smart Manufacturing.

The framework consists of three segments:

- **Core Foundation:** Those structural elements supporting operational excellence
- **Operations:** Actionable and measurable elements necessary for High Performance
- **Deployment/Delivery:** Smart Operations supporting Smart Manufacturing

Collectively, the model captures, analyzes, and delivers a full holistic decision support system. Note, this includes both human decision-making as well as online automated or AI decisions.

2.13.2 Core Foundation

This model is built on a structural base. As might be expected, the *Organizational Governance* model is the basis upon which successful operations management depends. Increasingly, adherence to a variety of *Standards* is required (in some cases by customer end users). This helps assure consistency of production as well as mitigating product liability issues—large component of risk management.

Smart Manufacturing requires different KSAs than currently possessed by traditional workforces and suppliers. A 2018 'Skills Gap Study' by the consulting firm Deloitte identified the following job description and associate KSAs for the Fourth Industrial Revolution.[119] These include:

- **Digital Twin Engineer:** Integrating across organizational departments/supply chain creating physical and digital solutions
- **Predictive Supply Network Analyst:** Use of digital tools including networking and data science
- **Robot Teaming Coordinator:** Optimizing the value of robots used in manufacturing
- **Digital Offering Manager:** The new product manager expanding digital offering in equipment sold
- **Drone Data Coordinator:** As appropriate supports field operations solution provided
- **Smart Factory Manager:** Drives Operational Excellence and Innovation at production facilities

- **Smart Scheduler:** Uses tools such as data analytics to drive improved schedule adherence
- **Smart Safety Supervisor:** Use of digital tools for enhanced EHS
- **UAM Flight Controller:** Use of digital tools for enhancing urban air mobility
- **Smart QA Manager:** Enhances overall productivity using digital tools to assure better quality assurance

Not every organization or factory will need all these individuals, i.e., airspace management may not be relevant for an automobile manufacturer. Finally, all these roles and their responsibilities marry IT with OT as shown previously. It will no longer be satisfactory for individuals such as supervisors, mechanical maintenance, etc. to remain in KSA silos. The combination of *hard* and *soft* (people) skills is a requirement.

A later Deloitte study indicated that in added that the required skills for the future workforce are largely nonexistent today. Moreover, like all sectors strong, effective DEI programs are vital for the manufacturing sector as well.[120]

Finally, *Tools* are defined as any devices whether software or hardware, i.e., wrenches, welding equipment, etc. necessary to accomplish process tasks and subtasks

2.13.3 OPERATIONS

This top vertical section of the framework lays out the execution or process and workflow required to operate a manufacturing facility in a sound manner. Read the framework from left to right Processes—Subtasks.

ISO 9001: 2015 defines the concept of Business *Process* as: 'A set of related or interacting activities, which transform inputs into outputs.' As such, tools such as Business Process Modeling (BPM) are useful for modeling this holistic configuration. Likewise, a *Procedure* is the specified way to carry out an activity or a process."[121]

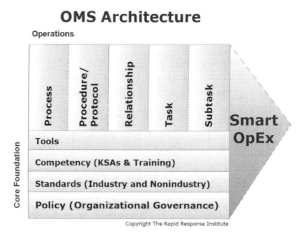

FIGURE 2.15   Operations management system governance architecture.

We can further define Procedures and a set of actions and the order they are to be carried out, whereas *Protocols* regarding said actions are a mandatory list of rules as well as behaviors required in their performance.[122] Increasingly, organizations are codifying operational behavior allowing little, if any, deviation on mission critical processes. In addition, the use of this type of system provides for an automatic audit with data captured one time as work is completed.

When the model refers to *Relationship*, it is referring to third-party suppliers and vendors who may be contractually obligated to complete work products. Frequently, a Bridging Document or similar instrument is included in a contract between the owner and third parties. Moreover, any subcontractors used are bound by the same agreement.

Tasks and associated *Subtasks* are part of the work or job to be undertaken. They are well defined and documented. In a manufacturing environment, there are few deviations that are acceptable unless engineering revision or other authorized changes are documented and incorporated into the procedure.

Finally, *Smart OpEx* aka Smart Operations is the output of the process. Keep in mind that each step in the process and all the tasks and subtasks have compliance and relevant policies and protocols built into them. This OMS framework does not require individual EHS or other activities to be documented separately.

It is important to understand that this type of OMS uses an underlying software workflow engine, and there is much more to the process than a simple set of software steps. Real-time data feed is often required as it can access historical and maintenance data and manuals. A sophisticated tool such as AI can access resulting data as well as update databases used by the OMS.

The incorporation of the framework will be discussed further in the following Roadmap section. A major goal of OMS is not to reduce complexity which will increase risk but enable humans to manage complicated challenges without being overwhelmed.

2.13.4 SAFETY AND ENVIRONMENTAL MANAGEMENT SYSTEM (SEMS)

As mentioned earlier, most organizations are incorporating Safety and Environmental Management into their organizations. Logically, SEMS must align and integrate with operations. Therefore, SEMS must be 'mapped' into OMS.

This is not a difficult process. Most SEMS models were developed by industry working groups and hence easily align with operational requirements.

2.13.5 THE LEARNING ORGANIZATION

The word 'Smart' suggests that this organizational transformation will be ongoing and sustained by new/updated processes and technologies. Organizations will be required to have a 'learning culture' and upskilling will be critical to continuous improvement. As with any change of this magnitude, Leadership from the top is essential.

The Organization Learning Framework depicted in the following figure is based on the 4I Organizational Learning model which can be defined as, "A conceptual framework of organizational learning process first proposed in 1999. The learning process consists of intuiting, interpreting, integrating, and institutionalizing within individual, group, and organization level."[123]

Briefly, the model consists of four processes as stated above with three levels, namely, individual, group, and organization. As shown in the framework graphic, Intuiting and Interpretation are individual behaviors. Interpretation flows over to the group level along with Integrating. Integrating and Institutionalizing are organizational-level activities.[124]

The model is straightforward and while originally posited as an instructor-led study before eLearning along with the various platforms available, i.e., mobility devices, it still works even for on-demand, self-study courses. It also covers third-party classes, courses, and even 'Train-the-Trainer' materials as well as internally developed curricula.

There are four major groupings:

- **Learners:** These individuals are at all levels in the firm and even its ecosystem, including top management
- **Learning Objectives and Syllabus (Course Outline):** What knowledge can the student expect to take away
- **Curriculum (Subjects) and Course Categories:** What a program might entail, i.e., AL software solution from a vendor
- **Enablers:** The set of tools and other materials as shown in the figure

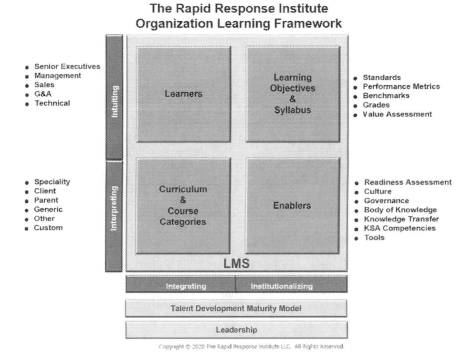

FIGURE 2.16   Organization learning framework.

The organization's Maturity Model will need to reflect 'Talent Development Maturity' as a component. Smart Manufacturing will require a broad set of KSAs and like any program of instruction, (adult) student skills and capabilities will vary.

Interestingly, as the airlines exited the self-imposed Covid-19 'Reduction In Force' (RIF), some pilots appear to have lost core flying knowledge. For example, a reactivated captain almost landed without lowering the landing gear.[125] If a highly competent and trained professional can make this basic mistake when furloughed for approximately 1 year, what exposure do other knowledge workers have when recalled after an industry downturn? In the Smart era, management should guard its talent wisely and not allow skill deterioration, regardless of the business cycle.

There is wealth of information and resources available. One that comes to mind is The Learning Guild, "a community of practice for those supporting the design, development, strategy, and management of organizational learning."[126] Other standards are available; for example, several ISO standards involve training.

2.13.6 Conclusion

There is a large Body of Knowledge both inside the manufacturing sector (discrete and process) as well as outside to draw upon. Moreover, IT has volumes on Best and Good Practices.

However, it is important to consider that what works well for one organization or sector may not work for another. The culture of an organization is an important consideration when 'cross pollinating' knowledge.

Learning from the successes and mistakes of those is the most inexpensive method to adopt new processes and technologies. Just do it with care and the proper due diligence.

2.14 ORGANIZATIONAL TRANSFORMATION PROCESS

Firms often enter the transformation process with trepidation. No doubt a major transformation such as Smart Manufacturing may be a journey and takes a great deal of time across all phases.

A Smart Manufacturing initiative is a project, and it needs to be treated as such. There is a substantial Body of Knowledge in this area, and we will refer to it in this section. We will focus on some 'lessons learned' as well as some things to keep an eye out for.

A Framework is provided to guide the process. In addition, the role of culture is explored as well as that of Change Management. As with any process, Smart Manufacturing must be measured. Ideas about KPIs to augment existing enterprise metrics are offered. Finally, we review the value Smart Manufacturing can have on the planet as well the construct of the Center of Excellence.

2.14.1 Career Opportunity?

A strong, well-qualified PM must be identified who will stay with the project. Preferably, this individual will be internal but is she/he if from a professional services firm or IT organization, *written* contractual assurances of the longevity of this individual is critical.

The PM should develop a project staff of the best and brightest people available. Organizations should attract these individuals as a major step in their career and track record should show they are well trained and given significant promotions when due.

Sometimes, especially operations and Profit and Loss individuals do not see appointment to an 'IT' project as a career enhancement. Since so many initiatives of this nature fail, many believe their career may actually be damaged and seek not to join or try to extract themselves as soon as possible.

Individuals at all levels both internal as well as external must see this project as Win—Win. Change Management is all about 'What's In It for Me?' If the project team cannot see personal value or does not believe the company line, the outcome is predictable.

2.14.2 IMPLEMENTATION

To paraphrase the real estate market, it is all about Execution, Execution, and Execution. In addition to the above comments, several major areas of risk have been identified. As with all lists herein, this is not all inclusive and some may not be applicable to some firms.

In no particular order, management should be aware of the following exposures. Mitigation strategies need to be incorporated into processes and even the governance model.

2.14.3 EXECUTIVE SPONSOR

A major manufacturing organization may spend hundreds of millions of dollars on their transformation to Smart Manufacturing. For those who doubt this statement, review the staggering costs of ERP implementations; by some accounts almost 200% over budget and almost 60% underperforming.[127]

These numbers are astounding and are confirmed by other sources quoted herein. So why such a high failure rate? Could be that the management announces Capital Investments and then 'assumes' middle management and a host of third-party providers will manage the process.

Smart expenditures are so enormous that CEOs cannot delegate Smart Manufacturing initiative responsibilities. Delegate a strong committed executive sponsor with authority and political clout needs to be in place and perhaps full time.

This appointment and commitment is the KEY to assure success with an acceptable economic value. Do not overlook this appointment. He/she needs to view this as a promotion opportunity. A commitment to be a future major player even at the industry sector level.

2.14.4 PROJECT MANAGEMENT OFFICE (PMO)

"A group or department within a business, government agency, or enterprise that defines and maintains standards for project management within the organization."[128] Large organizations may already have a PMO and it is likely that the Smart Manufacturing initiative would follow the policies and procedures defined therein. For those with no existing PMO, the initiative may set its own governance, keeping in mind the Best Practices of others.

2.14.5 THIRD-PARTY MANAGEMENT

Most organizations will need to employ professional services, software firms, and others. Even Fortune 500 firms do not have teams' waiting for projects. However, this dependence creates additional risks. Significant contractual constraints including Bridging Documents must be put in place. As little 'wiggle room' as possible.

For example, one of the worst things that can happen is when the (third-party) PM is transferred to another project early in the project. This can happen when the vendor wants to obtain another client. Moreover, hire the A Team and do not allow substitution with a B Team. It is also an example of 'bait and switch' which many firms do to obtain contracts.

2.14.6 SCOPE OF WORK

Defining what the project will accomplish along with the deliverables and time frame is critical. Things will change but going through this definition process will help management understand the issues and the project risk profile at a detailed level.

Many do not spend enough time and effort to get this statement correct, reasonable, and accomplishable. This sloppiness contributes to failure and disappointment.

Projects will have a Change Management Process. This process and authorization, price increases, etc. must be well defined and understood by all, including vendors, who may be called upon for rapid procurement. Blockchain may help in this regard.

2.14.7 START SMALL

A journey of a thousand steps begins with one, so the adage goes. This is true for projects and especially transformational projects. To alleviate concerns referenced above, it is useful to pick well managed, focused initial Pilot or Phase 1 with strong controls. The so-called 'low hanging fruit' should show success rapidly. This will be an enormous morale and internal selling point for the overall initiative.

Typically, there are several steps in project initiation. Starting small and working toward a global rollout, success can be more assured. These include:

- **Proof of Concept:** "A realization of a certain method or idea in order to demonstrate its feasibility, or a demonstration in principle with the aim of verifying that some concept or theory has practical potential."[129] Some may require this effort; however, if products and solutions are commercial, it is probably not necessary.
- **Pilot:** "An initial small-scale implementation that is used to prove the viability of a project idea."[130] Typically used with initiatives of this nature as a starting point.
- **Trial:** "Tests the implementation approach and its purpose is to manage the risk of implementation."[131] May or may not be used with this project.

- **Phases$_{1-n}$:** Large initiatives are usually divided into a series of phases. Partly, as stated above and partly the resource constraints of personnel, product delivery, and funding. Moreover, phasing is learning. What works and does not work in earlier phases can be applied to go forward.

While most do not want to accept the possibility, a phased approach allows the organization to end or STOP/REEVAUATE a global initiative at a number of steps. Sometimes, losses must be cut and there should be a recognition that 'this is just not working.' If a project is failing or needs to be reevaluated, stopping at a phase makes sense.

On the positive side, the phased model is controllable. Management can implement the entire (often multi-year) Smart Manufacturing with minimal, if any, impact on production.

2.14.8   RE-IMAGINE

Firms are often advised to re-imagine possibilities, most often by those consultancies and IT providers who stand to benefit. There is no question that Creative Destruction or disruption presents opportunities for nibble and/or startup firms.

It pays to assess where 'we are' and how we might 'leapfrog' the competition. The so-called 'thinking outside the box.' Disruptive change such as the digital camera or Uber dramatically and quickly changed their respective sectors. Management must search for those opportunities that will propel their companies.

2.14.9   FRAMEWORK

Any implementation process is two phased, initially led by the business needs. Phase II is focused on the technology and how it is deployed. The model is straightforward and note that there is transition built into it. At the beginning, business requirements are the sole focus, but as the process unfolds the technology required to enable Smart Manufacturing gradually becomes the emphasis.

The following figure depicts the six steps of this approach. Once the Smart Manufacturing transformational strategy is defined, the next five segments begin to address the tasks necessary to realize management's vision.

The final step, Business-Led Development is as it is named. All digitalization development programs must always keep the business needs in mind. Technology may be cool, but it is an enabler of the business. Executive sponsorship and the PM must make sure business needs are always front and center during transformation.

- **Business Strategy:** Identifying Key Business Strategies, Critical Success Factors, and Performance Metrics
- **Process Design:** Understanding the Key Processes and Work Practices Needed and the Underlying Organizational issues
- **Information Models:** Modeling the Structure, Content, and Information Flow Required to Effect the Processes
- **Systems and Technology:** Selection of the Required Application Systems, Technology Tools, and Architectures

- **Solutions Architecture:** Management of the Systems Integration Process and Technology Migration Issues
- **Business-Led Development:** Smart Manufacturing Implementation

This framework provides management with guidance and is part of the Roadmap to Smart Manufacturing. It lays out an approach organizations can follow or use to hold third-party implementation firms accountable to the Scope of Work and Deliverables against the agreed upon project timeline.

According to the PMI, over 70% of large-scale transformations fail. As one might expect, this risk is a major concern to senior management and perhaps even the Board of Directors. PMI provides training and Certification in the field of Organizational Transformation.[132]

Since this a 'bet your career' moment, it might make sense to avail oneself of this model and other knowledge available. Do not underestimate the complexity and breadth of this enterprise level transformation.

2.14.10   MODEL MAPPING

Throughout this chapter, we have posited a number of models. For example:

- Operational Excellence
- OMS
- SEMS
- Data
- Others such as Learning

High Level Framework Implementation Overview

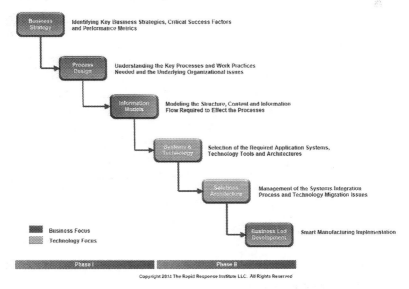

FIGURE 2.17   High-level framework implementation overview.

An organization can only have *one* business- and technology-enabled construct. People do not work with one model and then switch to another, for example, to file a report very well. All these constructs must be aligned, integrated, and have a common vocabulary and 'Single Sign On' to the enterprise Smart solution,

integrating into a single business model that employees as well as ecosystem personnel understand is critical for high performance. Moreover, it must be aligned with the organization's culture.

2.14.11 Culture

Every group of people, including commercial firms, has a culture, or set of common values and behaviors. It should not be surprising that each is different and often a source of pride and esprit de corps. As such, the interaction of cultures in the ecosystem can be complex and a source of value or derision.

While we often discuss organizational culture as a function of an *entity*, in reality it starts at the individual level. Ethnicity, background, and other variables define who a specific person is. That person is part of an organization which may consist of numerous groups each with their own (sub) culture.

Moreover, internal groups are constantly changing. Each acquisition and/or divestiture brings another separate culture that must be incorporated into the overall organization and its ecosystem which includes supply chain partners and their vendors as well as customers and perhaps their customers.

The term 'organizational culture' is often treated as if it is one homogeneous block. As shown in the following graphic, the level of cultural complexity is (C^n), in other words extremely complicated. It is very important that this issue be well understood before the transformation to Smart Manufacturing.

Most manufacturers produce in the communities in which their employees and others live. For example, environmental regulations must be adhered to and many support local schools and children's sports teams. In fact, many require STEM graduates for their future workforce. Each of these entities has a culture as well.

Culture is a source of high value, often couched as 'this is who we are.' A strong culture can differentiate an organization adding stakeholder value and better returns than competitors. Moreover, some cultures are more adaptable to change. Hence, the desire of the CEOs to be more Agile, Resilient, as well as Sustainable.

However, some, especially in highly regulated sectors, are not as amenable to rapid change. For example, Homeland Security has identified 16 Critical Infrastructure Sectors. Accordingly, these sectors, "whose assets, systems, and networks, whether physical or virtual, are considered so vital to the United States that their incapacitation or destruction would have a debilitating effect on security, national economic security, national public health or safety, or any combination thereof."[133]

Manufacturers in these sectors generally are genuinely concerned with safe operations, environmental impact, and safe consumption of their end products. As such, they are risk-averse, which translates into a slower uptake than others not so constrained.

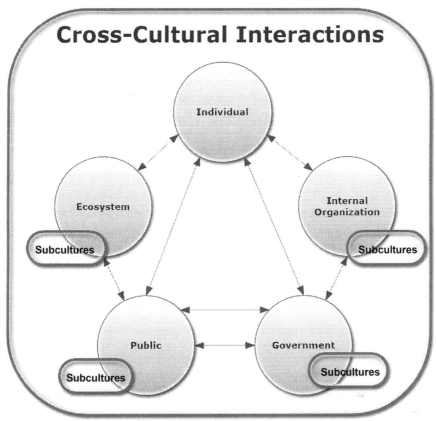

FIGURE 2.18   Cross-cultural interactions.

Often, third-party partners are retained for major initiatives. Likely, this will be the case for Smart Manufacturing initiatives as well. Management must retain suppliers that understands the organizational culture(s) and can work within said constraints.

Acknowledging culture and the behavior of individuals within it is one of THE most important action items in any transformation. Smart Manufacturing will be asking individuals to change their Behaviors both with humans and especially machines. The implementation of technology will change their situation or Conditions of daily life. The Relationships they have with others and their daily work will change as well. This RBC model is well understood and fundamental in transformation change models.

In addition, expect acculturation or the modification of cultures due to Smart Manufacturing. Management will need to take this into consideration going forward after the initial phases.

Moreover, in the Smart Manufacturing era, older hierarchical-based cultures as sometimes found in the traditional heavy industries may want to assess whether that approach is the best in the 21st century. For example, leader empathy is more effective than top-down authoritativeness.[134]

2.14.12 SAFETY

All organizations strive for safety and spend significant effort and funds training employees and contractors. Historically, the model has been along the lines of OSHA and its guidance and compliance requirements. This approach began to change over a decade ago when it became apparent that this approach was not meeting the needs of Critical Infrastructure sectors with their complexity and large ecosystems required to meet production and add to stakeholder value.[135]

The construct of a Safety Culture aka Culture of Safety roots are with the International Nuclear Safety Group (INSAG), which is under the auspices of the International Atomic Energy Agency (IAEA). It was first used in the 1986 "Summary Report on the Post-Accident Review Meeting on the Chernobyl Accident."[136] As noted herein, a strong Safety Culture is critical for a sustained and effective Smart Manufacturing firm and its ecosystem.

2.14.13 DIVERSE GROUPS

Large organizations have a diverse workforce, including its ecosystem. Collaboration on a regional and global scale involves a range of cultural diversity as shown in the previous figure.

Often, there are ethnic variations as well as gender, regional, and other aspects of cultural multiplicity that must be taken into consideration. This approach is consistent with ESG reporting requirements.

2.14.14 CHANGE MANAGEMENT MODEL

Change is always an issue with the deployment of enabling technologies and transformational changes in organizational processes and structure. Much has been written about this subject and processes are well understood. Still, to this day, up to 70% of change management initiatives fail.[137]

While there are many reasons, the one that leaps out is "buy-in." Management is often halfhearted and underfunds the process. Employees can sit back and wait for 'this to pass.'

The fundamental question that must be answered at all levels of the organizational ecosystem is, "What's In It for Me?"[138] If the individual from CEO down to the lowest level does not see a personal value proposition, then the likelihood of buy-in is low. Once this question is answered, change can be explosive.

For example, 2007 is seen as the birth date of the Smart Phone.[139] Prior to that, none of us knew we needed one. By some accounts, there are now approximately 6.4 billion phones worldwide.[140] Clearly, in less than 15 years, take-up has been phenomenal. The question has been answered by 6.4 (10^9) buyers. Even more if one assumes some have purchased multiple phones as technology advanced.

People change when they see the value to their lives. It can happen again.

2.14.15   R B C

The RBC Framework was originally posted as research for international business negotiations.[141] It has a more fundamental value in that is models basic human nature when interacting with others whether personally or at an organizational level. As such it is a *useful tool* when addressing transformation change.

If one sees the implementation of Smart Manufacturing as a Condition that is shifting, individual Behaviors will adjust accordingly. The resulting Relationships between individuals, groups, and even technology will be new.

Back to our smart phone metaphor, how many of us walk and drive while texting? A behavior and relationship with a communications device we never had before. This has even resulted in new regulations and laws, such as do not talk on the phone in a school zone.

The R B C construct provides those responsible for organizational change with a multi-dimensional model to help drive acceptance of innovative ideas quickly and with limited 'pushback.' As such it is part of the Smart Manufacturing Roadmap developed herein.

The OPM incorporating SEMS is an example where the structure is applicable. Structure and Process along with the need for Situational Awareness will cause behavioral changes, which is not always for the good. How many operational short-cuts are taken in an unsafe manner to meet a deadline or some other metric—see Deepwater Horizon et al.

2.14.16   Key Performance Indicators

One can argue that if it cannot be measured, it cannot be managed. Some posit that intangibles play a role in decision-making and this is most definitely part of the equation.[142] However, data-driven business model are in the preponderance.

Experienced executives will internalize data and make decisions incorporating knowledge and experience. That said, the data must be valid and reliable.

This is an issue Big Data, and its derivatives, are facing now. Examples of challenges include Covid-19 and climate change data sets.[143] Many are suspect, and the policy decisions made using them are suspect leading to an angry public discourse.

Likely, and for some time, Smart Manufacturing will have data issues. Management must put in place systems and not just technical expertise to assure human decision-making when it is necessary to override digital systems. As noted, a current example of technical decision-making failure is the Boeing Max 8 autopilot.[144]

When does a human override a computer will remain a major issue, likely for decades? A good set of KPIs can help mitigate the risk of poor or even bad decisions based on technology.

It is important to note that like any IT initiative, a Smart transformation must align with the core business process and culture of the organization.[145] If this alignment, including management's desire to dramatically change the organization is not understood and incorporated into this process, failure is much more likely.

The following KPIs are taken from a variety of public sources and are cited accordingly. Note that overall success using these indicators should be assessed and aligned with the Maturity Model previously described. They would be used in addition to existing organizational KPIs, not in 'lieu of.'

The above KPIs are not meant to be all inclusive and individual organizations may find others of greater value. Finally, any assessment of Smart Manufacturing metrics must be aligned with the way the business is run, i.e., ROIC.

TABLE 2.2
Key Performance Indicators

KPI	Description
Business sustainability (F)	Is enterprise risk mitigated? If so, can it be quantified?
Cloud application deployment (F)	New or 'Greenfield' applications developed using cloud or other new technology platforms
Combination of external and internal success metrics (F)	**External:** Measured by input from customers, i.e., feedback, net promoter score, etc. **Internal:** Measured by calibrations such as mean time to failure of variables, efficiency, etc.
Core value metric (F)	Digital transformation alignment with core metrics, values, and culture
Cost of user acquisition (F)	Economic cost of acquiring a new user
Customer experience (F)	The rate and/or success as customers navigate newly digitized organizational processes
Energy consumed per unit of manufacture/ production (A)	Measure of the amount and/or cost of energy required to produce one product or good in manufacturing from a total cost perp
Hours saved (F)	Number of (human) hours saved and may be an intangible such as efficiency or productivity. See EVPM
Operating expenses and contribution margin (F)	The ratio of the new operating cost (OpEx) the revenue stream after digitalization
Operational improvement (F)	Number of processes on 'new' software and impact, i.e., productivity, output increase, etc.
Outbound marketing performance (F)	If relevant, what is the return on investment from these initiatives?
Rate of innovation (F)	An excellent 'leading' indicator or the R&D cycle efficiency and ROI of digital investments
Return on capital employed (ROIC)	*A profitability ratio, measures how efficiently a company is using its capital to generate profits. The return on capital employed metric is considered one of the best profitability ratios*[146]
Revenue from new digital services (F)	Can new sources of revenue be attributed to smart technologies?
Safety and environmental management system (SEMS)	*The purpose of an SEMS is to enhance the safety and environmental performance of operations by reducing the frequency and severity of incidents*[147]

(Continued)

TABLE 2.2 (*Continued*)
Key Performance Indicators

KPI	Description
Team morale (F)	Impact of the digital transformation or cultural change as a result of the digital initiative
User lifetime value (F)	A function of user retention and the overall value received
Workforce productivity (F)	Bottom line value, i.e., increase in revenue per employee

(A)—Developed by the author.

(F)—"14 Important KPIs To Help You Track Your Digital Transformation." Forbes. https://www.forbes.com/sites/forbestechcouncil/2020/06/25/14-important-kpis-to-help-you-track-your-digital-transformation/?sh=53460ec49342. Accessed 1 Jun 2021.

2.14.17 DECARBONIZATION

According to one source, Smart digital technologies may accelerate global emission reduction up to 15% by 2030. Leveraging many of the technologies and associated processes herein, Smart Manufacturing and associated Smart Logistics supply chain can become more efficient reducing process cycle time as well as costs. Reducing energy consumption, minimizing 'rework,' greater precision, better quality assurance, and increased safety all add directly to a reduced carbon footprint.[148]

Each of these KPI or variables can be either directly or indirectly measured. The actual value to a given firm is thus documented and used as appropriate for financial management issues as well as components of ESG initiatives.

2.14.18 CENTER(S) OF EXCELLENCE

The journey to a fully Industry 4.0 organization will be long and never ending with new ideas, processes, product development as well as technology refresh. Organizations can depend on or third-party experts, in-house knowledge, or preferably both.

For example, most large organizations have an in-house legal team. Contracts, purchase order agreements, etc. can be handed by internal personnel (or offshore outsourcing) and this is probably the best and most cost-effective way to handle day to day legal issues.

Significant lawsuits require specific and often expensive expertise. Others such as patent law may need very narrow but deep knowledge.

There can be major problems when the firm outsources any key process or knowledge. Contractors, their subs, vendors, and others can create significant exposure for organizations if not handled well.

The adage, 'Just in Time inventory management works until manufacturing depends on someone's home garage' reflects a flaw in the system. When a critical partner or vender has performance issues or are acquired (perhaps by a competitor), disruptions to core processes are a potential risk.

Core Knowledge is competitive advantage. This Intellectual Property (IP) must be protected at all costs. Patents, Copyrights, and Trademarks etc. are good; however, Trade Secrets and Know How are perhaps most important.

One of the major problems with the first three types of IP is that organizations must publicize the said value. Trade Secrets on the other hand can be kept confidential. For example, the receipt for Coca Cola.

Developing an internal Center of Excellence (CoE) is a best practice assuring that the organization's commitment to Smart Manufacturing is protected and focused. This can also be an environment for testing new solutions such as vetting MVPs in a PoC process. Furthermore, educational processes and curriculum can be sponsored by a CoE.

Certain benefits include:

- **Resource Allocation:** Work with operations and manufacturing executives to assure proper skill set and human resource loading for a given project.
- **Linkage across Business Units and the Ecosystem:** Facilitate collaboration to solve problems and assure knowledge transfer.
- **Knowledge Repository and Best Practices:** Fundamental core knowledge is advanced.
- **Support HRO Processes:** Assure resources and knowledge is available to assure flexible production rapidly responding the issues that arise.
- **Development of Economic Value Investment Models:** Value is a fleeting and capturing is dynamic.

To achieve CoE success, there must be ongoing training, well-defined objectives, a predisposition to change (HRO), and investment in the effort.[149] The next section will define the execution model or construct.

2.14.19 Center of Excellence Construct

This Center of Excellence paradigm has evolved over several years and provides management with the tools necessary to develop and implement a Smart Manufacturing CoE. As shown in this figure, there are three components. At the center of the model is the Knowledge base of Good and Best Practices that will be managed per the above processes.

On the left is the Strategic Information Framework, which consists of all information at the Enterprise level as well as Smart Manufacturing and other data and associated processes. Moreover, governance, including IT, corporate, and ESG, is a major consumer of information developed by these systems.

On the right, the Change Management Framework consists of the following:

- **Expertise:** More than just skill sets, it includes the organization knowledge embodied internally as well in the ecosystem.
- **Information Technology:** Those systems providing data and information that feed the Strategic Information Framework systems.
- **Delivery:** Processes, tools supporting the CoE.

- **RR Matrix:** Fundamental business model of the firm.
- **Interdisciplinary Common Vocabulary:** Recognition that any organizational ecosystem must speak the same language. Not English vs. Spanish et al., but a common understanding of terminology. As noted, this is critical to an effective Safety Culture.

The intent is to capture, manage, and attain value from all the requirements to implement and sustain an effective Smart Manufacturing business model. Keep in mind, the model is dynamic and as change occurs it will be updated.

This basic framework provides management with an approach toward capturing and capitalizing on Smart Manufacturing. A Center of Excellence should be considered as an integral part of any initiative in this field. After all, a large organization will spend hundreds of millions if not more on this transformation. This investment must be managed, and ROI assured. Invest in a CoE.

Finally, there is a role for independent, third-party CoEs and industry groups. These can be valuable sources of knowledge and a venue to assure industry-wide conversations without Anti-Trust exposure. Industry organizations work closely with each member firm including Working Groups to advance issues including proposed regulations and their compliance process.

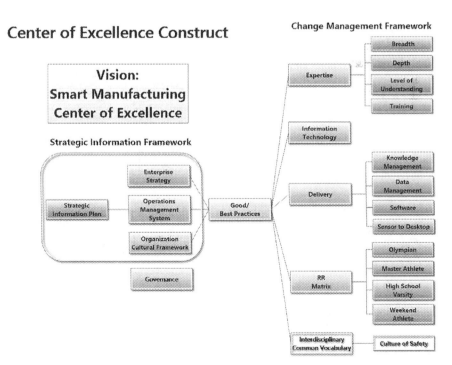

FIGURE 2.19 Smart manufacturing center of excellence construct.

One such group is the Centre for the Fourth Industrial Revolution, from the World Economic Forum, which is "a hub for global, multistakeholder co-operation to accelerate the benefits of science and technology."[150]

An investment in a Smart Manufacturing Center of Excellence will return shareholder value. This investment is highly recommended.

2.14.20 FINAL THOUGHTS

The transformational process is multi-dimensional. It is not simply deploying technology, but it does transfigure the organization into its future state. In this section, we discussed many of these dimensions and how they interact.

While complex and seemingly desperate, implementing a Smart Manufacturing initiative can be straightforward and not as intimidating as many believe. The high failure rate need not be *fait accompli*. This approach assures management that there is a codified solution to the transformation to the future.

2.15 ROADMAP FRAMEWORK

The global advisory firm, Ernst & Young (EY), reported that a CNBC survey found that in 2019 over $1.3 trillion was spent on digital transformation initiatives and approximately $900 billion was wasted as POCs, which were not commercially accepted.[151] To put this in perspective, the total 2020 revenue of Fortune's top three companies, Walmart, Amazon, and ExxonMobil, was approximately $1.07 trillion.[152]

This is an amazing figure and makes one wonder if it was a typographical error. How can organizations survive in a competitive world with such low performance? Most assuredly, not many promotions or bonuses were given for those in the $900 billion category.

The EY article went on to describe the usual suspects when IT projects fail; vision was unclear, unrealistic roadmaps, poor linkage to operational excellence programs, training and user experiences were unsatisfactory, and data quality and governance issues. Finally, cyber security concerns.

Moreover, McKinsey & Company is helping their clients address the issue of "Implementing 4IR at scale."[153] All the top professional services firms and major software providers are or soon will be offering Smart Manufacturing transformation. Others will offer products and services as well.

It is important that manufacturing firms become knowledgeable buyers who can perform appropriate due diligence and managerial oversight during the project or set of projects. This is not a program that can be outsourced but rather an interactive engagement among all those involved.

2.15.1 CURRENT STATUS

Cambridge Dictionary defines a Roadmap as, "a plan for how to achieve something: A business plan is a road map for achieving a vision or goal."[154] In other words, a defined process taking an organization from where it is to where it wants to be.

There is a substantial body of work regarding the development and use of Roadmaps. Most tend to be somewhat generic and generally fall into two categories: business and technology. These can be useful within the limitations of nonspecificity to Smart Manufacturing.

One can expect that better defined (fit for purpose) Roadmaps will emerge. For example, CESMII—The Smart Manufacturing Institute has developed Roadmap: 2017–2018.[155] CESMII indicates this program is ongoing as one would expect as Smart Manufacturing is taken up by all manufacturing industry sectors.

It is not the intent of this segment to replace or compete with the CESMII model or others. Rather, to posit a framework for the implementation of a Roadmap that meets the specific requirements of any given organization.

2.15.2 SMART MANUFACTURING MIND MAP

When an organization develops its Roadmap to become a Smart Manufacturer, it will include a high-level Timeline as well as a set of Tasks. In many ways, the Roadmap is a higher-level Project Plan. For purposes of this book whose readership comes from a broad population of international executives, technology managers, academia, and perhaps government, this would not be appropriate or even applicable.

The following figure depicts the full framework construct to implement SM. Simply tailor it to specific needs and/or use it as a quality control mechanism when using third-party tools. As with all sections of this chapter, it cannot be all inclusive due to the broad nature of the subject but is a 'go by' for developing and deploying Smart Manufacturing

The framework is divided into two categories: Primarily Economic and Primary Process. Keep in mind that linkages are extensive and it is not possible to show them all in this context.

It would be redundant to review each component depicted in the figure. Therefore, we will only address some additional thoughts to consider.

The points raised herein can pose a significant risk to the project as well as the sustainability of Smart Manufacturing. They are presented in this section to draw attention as the impact may be substantial.

2.15.3 MATURITY

One of the challenges Roadmaps or Project Plans are faced with is the maturity of both the Business Processes and the suite of Information and Operational Technologies. The following figure depicts Gartner's Magic Quadrant for Robotic Process Automation.[156] This suggests that major manufactures have few qualified vendors to pick from.

This is a useful tool to quickly assess available software suites. We recommend that manufacturers use this or other tools for *each* software application(s) under consideration.

Firms can then map applications to their needs. Inherent to this process is the Risk Profile associated with each software vendor relationship.

Smart Manufacturing Roadmap Framework

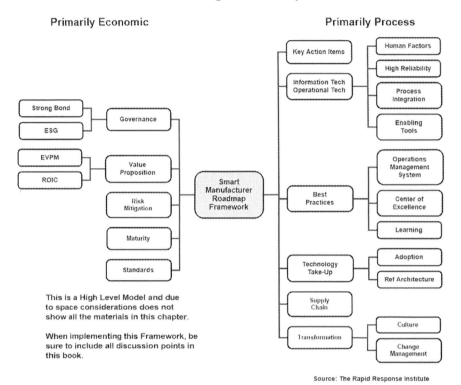

FIGURE 2.20   Smart manufacturing roadmap framework mind map.

Keep in mind that software application 'switching costs' are high. Personnel are trained in their use, data stores are connected as well as the integration with other applications (existing, new, and future).

This next figure is even more interesting. The research firm, Gartner, has published its Hype Cycle for Emerging Technologies in 202.[157] This is a useful view of technology and even the human change life cycle.

Some may be aware of the (circa 1969) Kübler-Ross Change Curve which appears the Hype Cycle is based on. It was initially developed to show the stages of significant (change) loss such as the death of a loved one.[158] These models are applicable for the transformation to Smart.

Many of the Information Technologies necessary for a Smart Manufacturing deployment do not appear to be very mature. This Risk must be incorporated into any decision process.

So, in addition to assessing the ability of a vendor to deliver, firms must also assess whether the applications desired/required will meet their needs. Again, Risk must be assessed for *each* application in the portfolio.

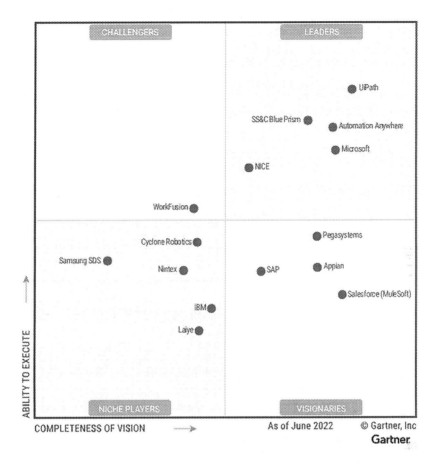

FIGURE 2.21 Magic quadrant for robotic process automation. (Gartner, Magic Quadrant™ for Robotic Process Automation, Saikat Ray, Arthur Villa et al., 25 July 2022. *Gartner and Magic Quadrant are registered trademarks of Gartner, Inc. and/or its affiliates in the U.S. and internationally and are used herein with permission. All rights reserved.*)

Moreover, the organization's Maturity and appetite for Risk will dictate the comfort level for this initiative. From the Technology Take-Up: Adoption Model, firms that consider themselves to be Early Majorities may be on the cusp for implementing Smart Manufacturing at the current time.[159]

Certainty, Late Majorities and Laggards may be very uncomfortable at present. One of the reasons why technology initiatives fail can be attributed to early adoption by firms whose cultures suggest they would be better off waiting.

One business case often made is, 'we must because everyone else is.' Words to this effect are frequently used when the IT in question is seen as cool.

We have made the case that sometimes Big Data analysis is not ready for prime time. Covid-19 may be a case in point. Surely, a source of much confusion and apparently inconsistent decision-making.

Hype Cycle for Emerging Technologies, 2022

FIGURE 2.22   Hype cycle for emerging technologies, 2022. (Gartner, Hype Cycle™ for Emerging Technologies, 2022, Melissa Davis, Gary Olliffe, 25 July 2022. *Gartner and Hype Cycle are registered trademarks of Gartner, Inc. and/or its affiliates in the U.S. and internationally and are used herein with permission. All rights reserved.*)

2.15.4   Filling the Skills Gap

A Learning Organization will have a continuing need for new skills. Not only new employees (experience or new graduate) will enter but technology and behavioral requirements will change.

This is another example of how the RBC framework works. The new Smart Conditions changes Behaviors resulting in different Relationships. Gradually, the organization's culture assimilates this value add.

While not directly addressed earlier, the maturity of the workforce must be considered as well. The advisory firm, Deloitte, has indicated that enhanced technical skills will need to be augmented with soft skills such as "creativity, leadership, critical thinking." [160]

Fortunately, organizations such as the Collaborative for Academic, Social, and Emotional Learning (CASEL) are advocating for Social and Emotional Learning (SEL) at the high school level.[161] This suggests that new additions to the workforce will have a level of ability with the soft skills.

The existing workforce can be taught SEL skills and training programs at all levels should add them to the curriculum. Keep in mind that there are three interrelated interactive fields to address: cognitive, social, and emotional.[162]

This is a wide range of expertise to build. It will be a continuing program throughout one's career and not a one-time or annual workshop. Moreover, organizational cultures must value SEL.

2.15.5 CONCLUSION

The Framework construct presented in this section is designed to help simplify this broad encompassing subject. A Smart Manufacturing Roadmap must retain the fidelity of the data, information, and processes within it. The high-level model presented is intended to help executives visualize the entire subject.

Major risks from technology were addressed in this section. Much of the technology 'hyped' for Smart Manufacturing is relatively immature. Moreover, not all vendors are equal in their ability to support and enable these initiatives.

Finally, hopefully the use of this Framework will reduce implementation and ongoing risk as organizations transform into more effective and safer production. As with any approach, use appropriate due diligence throughout this process.

Everything presented herein drives the Roadmap Framework. By following the guidelines developed herein, your risk of undertaking Smart Manufacturing should decrease and the value of your initiative should increase.

NOTES

1   "Smart Manufacturing." Gartner Glossary. https://www.gartner.com/en/information-technology/glossary/smart-manufacturing. Accessed 23 Aug 2021.
2   "Wisdom of Crowds." Investopedia. What is Wisdom of Crowds? https://www.investopedia.com/terms/w/wisdom-crowds.asp. Accessed 17 Sep 2021.
3   Venkataramani, Swetha. (2021). 6 Key Actions for a Successful Smart Manufacturing Strategy, Smarter With Gartner. https://www.gartner.com/smarterwithgartner/6-key-actions-for-a-successful-smart-manufacturing-strategy/. Accessed 23 Aug 2021.
4   "Suez Canal Blocked by Traffic Jam After Massive Container Ship Runs Aground." CNN. https://www.cnn.com/2021/03/24/middleeast/suez-canal-container-ship-intl-hnk/index.html. Accessed 30 Mar 2021.
5   "Rapid Response Management: Thriving in the New World Order." The Rapid Response Institute. Changing the Dialogue. https://therrinstitute.com/wp-content/uploads/2017/10/rapid_response_management_-_thriving_in_the_new_world_order_2.0_-_january_2009.pdf. Accessed 23 Sep 2021.
6   Shemwell, Scott M. (2011). *Essays on Business and Information II: Maximizing Business Performance*. New York: Xlibris, p. 13.
7   "Human Factors and Ergonomics." Wikipedia. https://en.wikipedia.org/wiki/Human_factors_and_ergonomics. Accessed 7 Sep 2021.
8   "High Reliability." Patient Safety Network (PSNet). https://psnet.ahrq.gov/primer/high-reliability. Accessed 7 Sep 2021.
9   "Human Systems Integration." Undersecretary of Defense for Research and Engineering; Department of Defense. https://ac.cto.mil/hsi/. Accessed 7 Sep 2021.
10 "Securing America's Critical Energy Infrastructure: 5 Steps for Operators." Pipeline & Gas Journal. https://pgjonline.com/magazine/2021/september-2021-vol-236-no-9/features/securing-america-s-critical-energy-infrastructure-5-steps-for-operators. Accessed 18 Sep 2021.
11 "Human–Machine Interface in the Age of Digitalization: Can the Machine be Trusted and When Should the Human Intervene?" The Rapid Response Institute. https://therrinstitute.com/human-machine-interface-in-the-age-of-digitalization-can-the-machine-be-trusted-and-when-should-the-human-intervene/. Accessed 7 Jan 2021.

12 "Situation Awareness." Wikipedia. Wikimedia Foundation. https://en.wikipedia. org/wiki/Situation_awareness#:~:text=Situational%20awareness%20or%20situation%20awareness,projection%20of%20their%20future%20status. Accessed 7 Jan 2021.

13 "Human Error: 8 Eye-Popping Sets of Stats and Examples." Ocrolus. https:// www.ocrolus.com/blog/human-error-8-eye-popping-sets-of-stats-and-examples/. Accessed 7 Sep 2021.

14 "Lessons Learnt from Alarm Management in a Combined-Cycle Gas Turbine Power Plant." Science Direct. https://www.sciencedirect.com/science/article/abs/ pii/B9780444639653504128. Accessed 18 Sep 2021.

15 "What is Ergonomics?" International Ergonomics Association. Definition and Applications. https://iea.cc/what-is-ergonomics/. Accessed 22 Sep 2021.

16 "Human-Factors Engineering Bioengineering." Britannica. https://www.britannica. com/topic/human-factors-engineering. Accessed 22 Sep 2021.

17 "Human Systems Integration." SEBoK: Guide to the Systems Engineering Body of Knowledge. https://www.sebokwiki.org/wiki/Human_Systems_Integration#:~:text= Human%20Systems%20Integration.%20Human%20systems%20integration%20 %28HSI%29%20is,elements%2C%20an%20essential%20enabler%20to%20systems%20engineering%20practice. Accessed 22 Sep 2021.

18 Ibid.

19 "Operationalizing Normal Accident Theory for Safety-Related Computer Systems." ScienceDirect. https://www.sciencedirect.com/science/article/abs/pii/ S0925753505000743#:~:text=Normal%20accident%20theory%20(NAT)%20 explains%20that%20some%20system%20accidents%20are,remains%20theoretical%20rather%20than%20empirical. Accessed 18 Sep 2021.

20 Weick et al. (2008). *Organizing for High Reliability: Processes of Collective Mindfulness.* California: Sage.

21 Holland, Winford "Dutch" E. and Shemwell, Scott M. (2014). *Implementing a Culture of Safety: A Roadmap to Performance-Based Compliance.* New York: Xlibris.

22 "Digitalization in the Service Supply Chain – Making the Investment Case." OnProcess. https://www.onprocess.com/blog/digitalization-in-the-service-supply-chain-making-the-investment-case/. Accessed 2 Sep 2021.

23 "The Top 9 reasons for IT Project Failure: Is Your Project at Risk?" atSpoke. https:// www.atspoke.com/blog/it/reasons-for-it-project-failure/. Accessed 19 Sep 2021.

24 "The ABCs of the Critical Path Method." Harvard Business Review. https://hbr. org/1963/09/the-abcs-of-the-critical-path-method. Accessed 19 Sep 2021.

25 Shemwell, Scott M. (1997, September). The Economic Value of Timely Information and Knowledge, Key to Business Process Integration Across Boundaries in the Oil & Gas Extended Value Chain. *Proceedings of the Gulf Publishing 3rd International Conference and Exhibition on Exploration & Production Information Management.* Houston.

26 "A Windy Position." The Rapid Response Institute. https://therrinstitute.com/a-windy-position/. Accessed 18 Sep 2022. Key OE Processes and Enabling

27 "Environmental, Social, and Governance (ESG) Criteria." Investopedia. https:// www.investopedia.com/terms/e/environmental-social-and-governance-esg-criteria.asp. Access 2 Apr 2021.

28 Shemwell, Scott M. (1993). Management Theory – Evolution Not Revolution, *Proceedings of the 11th Annual Conference of the Association of Management,* 11 (2), pp. 74–78.

29 "Agency Theory." Investopedia. https://www.investopedia.com/terms/a/agencytheory.asp. Accessed 2 Apr 2021.
30 "Maturity Model." Wikipedia. https://en.wikipedia.org/wiki/Maturity_model. Accessed 19 Sep 2021.
31 "Build Capability, Drive Performance." ISACA. https://cmmiinstitute.com/. Accessed 19 Sep 2021.
32 "Maturity Model." Wikipedia. https://en.wikipedia.org/wiki/Maturity_model. Accessed 19 Sep 2021.
33 "Development of a Digitalization Maturity Model for the Manufacturing Sector." IEEE. https://ieeexplore.ieee.org/document/8436292. Accessed 24 Sep 2021.
34 "OMG Mission & Vision." OMG Standard Development Organization. https://www.omg.org/about/index.htm. Accessed 20 Sep 2021.
35 "The Assessment Process." Missouri State. https://www.missouristate.edu/assessment/the-assessment-process.htm. Accessed 20 Sep 2021.
36 "Asset/Equipment Integrity Governance: Operations–Enterprise Alignment." The Rapid Response Institute. https://therrinstitute.com/wp-content/uploads/2017/10/asset_integrity_governance_-ver_1.1.pdf. Accessed 20 Sep 2021.
37 "Safety Management Systems Standards and Guidelines: A Comparative Analysis." American Society of Safety Professionals. Safety Management: Peer-Reviewed. https://www.assp.org/docs/default-source/psj-articles/f2shekari_0920.pdf?sfvrsn=86db8847_2. Accessed 24 Sep 2021.
38 "A Review of Technology Acceptance and Adoption Models and Theories." ScienceDirect. https://www.sciencedirect.com/science/article/pii/S2351978918304335. Accessed 20 Sep 2021.
39 "The 5 Customer Segments of Technology Adoption." On Digital Marketing. https://ondigitalmarketing.com/learn/odm/foundations/5-customer-segments-technology-adoption/. Accessed 22 Sep 2021.
40 "The Calculus of Value." Executive Briefing. https://secureservercdn.net/198.71.233.111/c3c.2cb.myftpupload.com/wp-content/uploads/2020/01/The-Calculus-of-Value.pdf. Accessed 21 Sep 2021.
41 Ibid.
42 "We Deliver Actionable, Objective Insight to Executives and Their Teams." Gartner. https://www.gartner.com/en/about. Accessed 21 Sep 2021.
43 "Positioning technology players within a specific market." Gartner. Gartner Magic Quadrant. https://www.gartner.com/en/research/methodologies/magic-quadrants-research. Accessed 21 Sep 2021.
44 "Gartner Hype Cycle." Gartner. https://www.gartner.com/en/research/methodologies/gartner-hype-cycle. Accessed 21 Sep 2021.
45 "Gartner Magic Quadrant." Gartner. https://www.gartner.com/en/research/methodologies/magic-quadrants-research. Accessed 18 Sep 2022.
46 "Magic Quadrant for Robotic Process Automation." Gartner. https://www.gartner.com/doc/reprints?id=1–26QWQ67B&ct=210709&st=sb. Accessed 21 Sep 2021.
47 "Gartner Magic Quadrant & Critical Capabilities." Gartner. https://www.gartner.com/en/research/magic-quadrant. Accessed 21 Sep 2021.
48 "Gartner Hype Cycle." Gartner. https://www.gartner.com/en/research/methodologies/gartner-hype-cycle. Accessed 21 Sep 2021.
49 "ISA95, Enterprise-Control System Integration." International Society of Automation (ISA). https://www.isa.org/standards-and-publications/isa-standards/isa-standards-committees/isa95. Accessed 24 Sep 2021.

50 "Is the Purdue Model Still Relevant?" AutomationWorld. https://www.automa-tionworld.com/factory/iiot/article/21132891/is-the-purdue-model-still-relevant. Accessed 24 Sep 2021.

51 "Hype Cycle for Enterprise Architecture, 2022." Gartner. https://www.gartner.com/account/signin?method=initialize&TARGET=http%3A%2F%2Fwww.gartner.com%2Fdocument%2F4015847%3Fref%3DTypeAheadSearch%26toggle%3D1. Accessed 18 Sep 2022.

52 "Attaining & Sustaining Operational Excellence: A Best Practice Implementation Model." The Rapid Response Institute. https://secureservercdn.net/198.71.233.111/c3c.2cb.myftpupload.com/wp-content/uploads/2017/12/OE-Best-Practice-Implementaion-2017.pdf. Accessed 25 Sep 2021.

53 "Let's Play A Game." The Rapid Response Institute. https://therrinstitute.com/lets-play-a-game/. Accessed 29 Sep 2021.

54 "Operational Excellence Management System." Chevron. https://www.chevron.com/-/media/shared-media/documents/oems_overview.pdf. Accessed 29 Sep 2021.

55 Holland, Winford "Dutch" E. and Shemwell, Scott M. (2014). *Implementing a Culture of Safety: A Roadmap to Performance-Based Compliance.* New York: Xlibris.

56 "Asset/Equipment Integrity Governance: Operations–Enterprise Alignment." The Rapid Response Institute. Changing the Dialogue. https://therrinstitute.com/wp-content/uploads/2017/10/asset_integrity_governance_-ver_1.1.pdf. Accessed 29 Sep 2021.

57 "What is a Lakehouse?" Databricks. https://databricks.com/blog/2020/01/30/what-is-a-data-lakehouse. html. Accessed 29 Sep 2021.

58 "Data for Climate Action: From Covid to the Climate." Open data institute. https://theodi.org/article/data-for-climate-action-i-from-covid-to-the-climate/. Accessed 1 Sep 2021.

59 "Understanding the Evolution of Standards: Alignment and Reconfiguration in Standards Development and Implementation Arenas." http://homepages.ed.ac.uk/grahami/network/papers/Edinburgh%20-%20standards-EASST2004.pdf. Accessed 9 Sep 2021.

60 "There is Always More to Learn and Sometimes We Learn That We Do not Have it Quite Right." https://scienceisneversettled.com/. Accessed 9 Sep 2021.

61 "A Brief History of Standards." Ripple. https://ripple.com/insights/a-brief-history-of-standards/. Accessed 9 Sep 2021.

62 Shemwell, Scott M. (1993). Management Theory – Evolution Not Revolution, *Proceedings of the 11th Annual Conference of the Association of Management*, 11 (2), pp. 74–78.

63 "Information Technology Standards." ANSI Webstore. https://webstore.ansi.org/industry/information-technology/standards. Accessed 23 Aug 2021.

64 "Current Standards Landscape for Smart Manufacturing Systems." National Institute of Standards and Technology. https://nvlpubs.nist.gov/nistpubs/ir/2016/NIST.IR.8107.pdf. Accessed 24 Sep 2021.

65 "ISO/IEC TR 63306-1:2020 Smart Manufacturing Standards Map (SM2) – Part 1: Framework." ISO. https://www.iso.org/standard/81277.html. Accessed 24 Sep 2021.

66 "MTConnect Standardizes Factory Device Data." MTConnect. https://www.mtcon-nect.org/. Accessed 24 Aug 2021.

67 "Pilot/Controller Glossary." FAA. https://www.faa.gov/air_traffic/publications/media/pcg_4-03-14.pdf. Accessed 24 Sep 2021.

68 "Demystifying the Digital Thread and Digital Twin Concepts." Industry Week. The Digital Thread. https://www.industryweek.com/technology-and-iiot/systems-integration/article/22007865/demystifying-the-digital-thread-and-digital-twin-concepts. Accessed 7 Sep 2021.

69 "Why Having the Right Data at the Right Time is Paramount to Corporate Legal's New Normal." Law.com. https://www.law.com/corpcounsel/2020/08/25/why-having-the-right-data-at-the-right-time-is-paramount-to-corporate-legals-new-normal/?slreturn=20210807195734. Accessed 7 Sep 2021.

70 "How Does a Digital Twin Work?" IBM. https://www.ibm.com/topics/what-is-a-digital-twin. Accessed 7 Sep 2021.

71 "Blockchain." Wikipedia. https://en.wikipedia.org/wiki/Blockchain. Accessed 22 Sep 2021.

72 "Blockchain in Manufacturing." IBM. Enabling and trust and automation across boundaries. https://www.ibm.com/blockchain/industries/manufacturing. Accessed 23 Sep 2021.

73 "Global Standards Mapping Initiative (GSMI)." Global Blockchain Business Council. GBBC. https://gbbcouncil.org/gsmi/#:~:text=Global%20Standards%20Mapping%20Initiative%20%28GSMI%29%20The%20Global%20Blockchain,to%20survey%20blockchain%20standards.%20We%20Mapped%20Data%20From%3A. Accessed 24 Sep 2021.

74 "Major Industrial Accidents: The Reasons and the Reactions. Tennessee Research and Creative Exchange (TRACE)." The University of Tennessee. https://trace.tennessee.edu/cgi/viewcontent.cgi?article=1410&context=utk_chanhonoproj. Accessed 27 Sep 2021.

75 "Operational Complexity: Risk Model Insufficiency." The Rapid Response Institute. https://therrinstitute.com/operational-complexity-risk-model-insufficiency/. Accessed 27 Sep 2021.

76 "Bowtie Analysis and Barrier-Based Risk Management." Pharmaceutical Engineering. ISPE. https://ispe.org/pharmaceutical-engineering/january-february-2018/bowtie-analysis-and-barrier-based-risk-management. Accessed 27 Sep 2021.

77 "Asset/Equipment Integrity Governance: Operations–Enterprise Alignment." The Rapid Response Institute. Changing the Dialogue. https://therrinstitute.com/wp-content/uploads/2017/10/asset_integrity_governance_-ver_1.1.pdf. Accessed 27 Sep 2021.

78 "How Dare You!!" The Rapid Response Institute. Critical Mass Blog. https://therrinstitute.com/how-dare-you/. Accessed 7 Sep 2021.

79 "The Top 10 IT Disasters of All Time." SDNet. https://www.zdnet.com/article/the-top-10-it-disasters-of-all-time-5000177729/. Accessed 7 Sep 2021.

80 "Interpretation: Disclosure of Year 2000 Issues and Consequences by Public Companies, Investment Advisers, Investment Companies, and Municipal Securities Issuers." Securities and Exchange Commission. https://www.sec.gov/rules/interp/33-7558.htm. Accessed 7 Sep 2021.

81 "Sources of Systemic Error or Bias: Information Bias.' ERIC Notebook. https://sph.unc.edu/wp-content/uploads/sites/112/2015/07/nciph_ERIC14.pdf. Accessed 7 Sep 2021.

82 "Selection Bias." Wikipedia. https://en.wikipedia.org/wiki/Selection_bias. Accessed 7 Sep 2021.

83 "Confounding." Catalogue of Bias. https://catalogofbias.org/biases/confounding/. Accessed 7 Sep 2021.

84 "Cascading Failure. Wikipedia. https://en.wikipedia.org/wiki/Cascading_failure. Accessed 7 Sep 2021.

85 "Data Bias: The Latent or Unobserved." The Rapid Response Institute. Critical Mass Blog. https://therrinstitute.com/data-bias-the-latent-or-unobserved/. Accessed 7 Sep 2021.

86 "Data Bias and What it Means for Your Machine Learning Models." Explorium. https://www.explorium.ai/blog/data-bias-and-what-it-means-for-your-machine-learning-models/. Accessed 7 Sep 2021.

87 "Corporate Overview." MITRE. https://www.mitre.org/about/our-history. Accessed 18 Sep 2021.

88 "Enterprise Matrix." MITRE. https://attack.mitre.org/matrices/enterprise/. Accessed 18 Sep 2021.

89 "Mobile Matrices." MITRE. https://attack.mitre.org/matrices/mobile/. Accessed 18 Sep 2021.

90 "ATT&CK® for Industrial Control Systems." MITRE Partner Network. https://collaborate.mitre.org/attackics/index.php/Main_Page. Accessed 18 Sep 2021.

91 "The Colonial Pipeline Ransomware Hackers had a Secret Weapon: Self-Promoting Cybersecurity Firms." MIT Technology Review. Computing/Cybersecurity. https://www.technologyreview.com/2021/05/24/1025195/colonial-pipeline-ransomware-bitdefender/. Accessed 18 Sep 2021.

92 "Cybersecurity Standards." Wikipedia. https://en.wikipedia.org/wiki/Cyber security_standards. Accessed 18 Sep 2021.

93 "Global Shortage of Computer Chips Hits US Manufacturing." wttw. https://news.wttw.com/2021/07/29/global-shortage-computer-chips-hits-us-manufacturing. Accessed 7 Sep 2021.

94 "Punched Card." Wikipedia. https://en.wikipedia.org/wiki/Punched_card. Accessed 10 Sep 2021.

95 "Guidance on Enterprise Risk Management." Committee of Sponsoring Organizations' (COSO). https://www.coso.org/pages/erm.aspx. Accessed 22 Sep 2021.

96 "Systemic Risk vs. Systematic Risk: What's the Difference?" Investopedia. https://www.investopedia.com/ask/answers/09/systemic-systematic-risk.asp. Accessed 22 Sep 2021.

97 "Examining and Learning from Complex Systems Failures." Uptime Institute. https://journal.uptimeinstitute.com/examining-and-learning-from-complex-systems-failures/. Accessed 22 Sep 2021.

98 Boebert, Earl and Blossom, James M. (2017). *Deepwater Horizon: A System Analysis of the Macondo Disaster.* Cambridge, MA: Harvard University Press..

99 "When Big Data Goes Bad." Fortune. https://fortune.com/2013/11/05/when-big-data-goes-bad/. Accessed 11 Oct 2021.

100 "In Defense of Humans – Machines are Not Ready Yet." The Rapid Response Institute. https://therrinstitute.com/in-defense-of-humans-machines-are-not-ready-yet/. Accessed 11 Oct 2021.

101 "The Facebook Whistleblower Says its Algorithms are Dangerous. Here's why." MIT Technology Review. https://www-technologyreview-com.cdn.ampproject.org/c/s/www.technologyreview.com/2021/10/05/1036519/facebook-whistleblower-frances-haugen-algorithms/amp/. Accessed 11 Oct 2021.

102 "Has AI Found a New Foundation?" The Gradient. https://thegradient.pub/has-ai-found-a-new-foundation/?fbclid=IwAR1JuIzB87PlVh2mpwcxugJslOgc5Yh4S vSQ-ZkONCLaEvjgtR8YTSp15C4. Accessed 13 Sep 2021.

103 "How Automation Increases the Effectiveness of an Organization." Robotics & Automation News. https://roboticsandautomationnews.com/2021/01/14/how-

automation-increases-the-effectiveness-of-an-organization/39540/. Accessed 17 Sep 2021.

104 Ibid.

105 "A Guide to the Automation Body of Knowledge, Third Edition." International Society of Automation. https://www.isa.org/products/a-guide-to-the-automation-body-of-knowledge-th-1. Accessed 17 Sep 2021.

106 "Customer Relationship Management." Wikipedia. https://en.wikipedia.org/wiki/Customer_relationship_management. Accessed 17 Sep 2021.

107 "Asset Integrity Management." DNV. https://www.dnv.com/services/asset-integrity-management-1380#. Accessed 18 Sep 2021.

108 Shemwell, Scott M. (2012, March). *Integrity Management: Issues & Trends Facing the 21st Century Energy Industry.* The Rapid Response Institute.

109 "Enhancing Factories Through Digital-First Manufacturing." Industry today. https://industrytoday.com/enhancing-factories-through-digital-first-manufacturing/. Accessed 27 Sep 2021.

110 "IDC Reveals 2021 Worldwide Digital Transformation Predictions; 65% of Global GDP Digitalized by 2022, Driving Over $6.8 Trillion of Direct DX Investments from 2020 to 2023." IDC. https://www.idc.com/getdoc.jsp?containerId=prUS46967420. Accessed 19 Sep 2021.

111 Shemwell, Scott M. (2004, November). Governing Value. Governing Energy. Executive Briefing – Business Value from Technology. Vol 3 No. 11.

112 "Pricing and Digitalization: Why Digitalization is a Boardroom Topic." Simon-Kucher. https://www.simon-kucher.com/en-us/blog/pricing-and-digitalization-why-digitalization-boardroom-topic. Accessed 30 Sep 2021.

113 "Identifying Value at Stake for Society and Industry." World Economic Forum. http://reports.weforum.org/digital-transformation/identifying-value-at-stake-for-society-and-industry/. Accessed 30 Sep 2021.

114 "Enterprise of Value: Governance of IT Investments." IT Governance Institute. The Val IT Framework 2.0. https://csbweb01.uncw.edu/people/ivancevichd/classes/MSA%20516/Extra%20Readings%20on%20Topics/IS%20Governance/IT%20Value/Val%20IT.pdf. Accessed 30 Sep 2021.

115 "Economic Value Proposition Matrix." The Rapid Response Institute. https://therrinstitute.com/evpm-welcome/. Accessed 30 Sep 2021.

116 Niven, Chris and Prouty, Kevin. (2017). *IDC TechBrief: Operations Management Systems.* IDC.

117 "Operations Management Systems: Stripping Out The Complexity." PEX Network. https://www.processexcellencenetwork.com/business-process-management-bpm/articles/operations-management-systems-stripping-out-the-co. Accessed 4 Sep 2021.

118 "Operations Management System." Kinder Morgan. https://www.kindermorgan.com/About-Us/OMS. Accessed 4 Sep 2021.

119 "2018 Skills Gap in Manufacturing Study." Deloitte. https://www2.deloitte.com/us/en/pages/manufacturing/articles/future-of-manufacturing-skills-gap-study.html. Accessed 4 Sep 2021.

120 "Creating Pathways for Tomorrow's Workforce Today: Beyond Reskilling in Manufacturing." Deloitte. https://www2.deloitte.com/us/en/insights/industry/manufacturing/manufacturing-industry-diversity.html. Accessed 4 Sep 2021.

121 "Differences between Processes, Procedures and Work Instructions." LinkedIn. https://www.linkedin.com/pulse/differences-between-processes-procedures-work-instructions-bpm. Accessed 4 Sep 2021.

122 "What is the Difference between Protocols and Procedures?" HiNative. https://hinative.com/en-US/questions/231789. Accessed 4 Sep 2021.

123 "What is 4I Organizational Learning Model." IGI Global. https://www.igi-global.com/dictionary/4i-organizational-learning-model/58149. Accessed 30 Sep 2021.

124 "An Organizational Learning Framework: From Intuition to Institution." Academy of Management Review. https://cmapspublic.ihmc.us/rid%3D1222355636953_663250744_13307/Organizational%2520Learning%2520Framework%2520From%2520Intuition%2520to%2520Institution.pdf. Accessed 30 Sep 2021.

125 "Out-of-Practice Airline Pilots Are Making Errors Back in the Air." Bloomberg. https://www.bloomberg.com/news/features/2021-10-14/plane-crash-risks-rise-as-covid-19-leaves-airline-pilots-out-of-practice. Accessed 19 Oct 2021.

126 "About The Learning Guild." The Learning Guild. https://www.learningguild.com/content/54/about-the-guild/. Accessed 30 Sep 2021.

127 "Cost and Schedule Overruns – Functionality Under-runs." ResearchGate. https://www.researchgate.net/figure/Cost-and-Schedule-overruns-Functionality-under-runs_fig3_237305057. Accessed 30 Sep 2021.

128 "Project Management Office." Wikipedia. https://en.wikipedia.org/wiki/Project_management_office. Accessed 30 Sep 2021.

129 "Proof of Concept." Wikipedia. https://en.wikipedia.org/wiki/Proof_of_concept. Accessed 30 Sep 2021.

130 "Definitions." Association of Project Management (APM). https://www.apm.org.uk/resources/find-a-resource/what-is-the-difference-between-a-trial-and-a-pilot/. Accessed 30 Sep 2021.

131 Ibid.

132 "Upskill with the Organizational Transformation Series." Organizational Transformation Series. Project Management Institute. https://www.pmi.org/organizational-transformation. Accessed 30 Sep 2021.

133 "Critical Infrastructure Sectors." Cybersecurity & Infrastructure Security Agency. https://www.cisa.gov/critical-infrastructure-sectors. Accessed 31 Aug 2021.

134 Proof That Positive Work Cultures Are More Productive." Harvard Business Review. https://hbr-org.cdn.ampproject.org/c/s/hbr.org/amp/2015/12/proof-that-positive-work-cultures-are-more-productive. Accessed 20 Sep 2021.

135 "Safety Culture." Wikipedia. https://en.wikipedia.org/wiki/Safety_culture. Accessed 18 Sep 2021.

136 "Summary Report on the Post-accident Review Meeting on the Chernobyl Accident." International Nuclear Safety Advisory Group. International Atomic Energy Agency (IAEA). https://www.iaea.org/publications/3598/summary-report-on-the-post-accident-review-meeting-on-the-chernobyl-accident. Accessed 20 Sep 2021.

137 "1 Reason Why Most Change Management Efforts Fail." Forbes. https://www.forbes.com/sites/brentgleeson/2017/07/25/1-reason-why-most-change-management-efforts-fail/?sh=837dead546b7. Accessed 31 Aug 2021.

138 "I Hate These Things: Why Does This Always Happen to Me?" The Rapid Response Institute. https://therrinstitute.com/covid-19-business-continuity-resources/. Accessed 31 Aug 2021.

139 "A Brief History of the Smartphone." ScienceNode. https://sciencenode.org/feature/How%20did%20smartphones%20evolve.php. Accessed 31 Aug 2021.

140 "Number of Smartphone Users from 2016 to 2021." Statista. https://www.statista.com/statistics/330695/number-of-smartphone-users-worldwide/. Accessed 31 Aug 2021.

141 Weiss, Stephen E. (1993, May). Analysis of Complex Negotiations in International Business: The RBC Perspective*. *Organization Science*. Vol. 4, No. 2.

142 "Myth: If You Can't Measure It, You Can't Manage It." The W. Edwards Deming Institute. https://deming.org/myth-if-you-cant-measure-it-you-cant-manage-it/. Accessed 31 Aug 2021.

143 "COVID-19, Data, and Climate Change." RMI. https://rmi.org/covid-19-reinforces-datas-critical-intervention-role-with-climate-change/. Accessed 31 Aug 2021.

144 "Linear Metrics in Non-Linear Times?" The Rapid Response Institute. https://therrinstitute.com/linear-metrics-in-non-linear-times/. Accessed 31 Aug 2021.

145 "Workflow Alignment. The Rapid Response Institute. " https://secureservercdn.net/198.71.233.111/c3c.2cb.myftpupload.com/wp-content/uploads/2018/10/Workflow-Alignment-October-6–2014.pdf. Accessed 1 Sep 2021.

146 "What is Return on Capital Employed?" CFI. https://corporatefinanceinstitute.com/resources/knowledge/finance/return-on-capital-employed-roce/. Accessed 23 Sep 2021.

147 "Safety and Environmental Management Systems – SEMS." Bureau of Safety and Environmental Enforcement (BSEE). https://www.bsee.gov/reporting-and-prevention/safety-and-environmental-management-systems. Accessed 24 Sep 2021.

148 "Solving the Climate Crisis with Industry 4.0 (Reader Forum)." Enterprise iot insights. https://enterpriseiotinsights.com/20200709/channels/reader-forum/solving-the-climate-crisis-with-industry-4-0-reader-forum. Accessed 23 Sep 2021.

149 "Centers of Excellence, What Are & Why Are They Necessary?" Nexus Integra. Digital transformation. https://nexusintegra.io/centers-of-excellence-coe/. Accessed 24 Sep 2021.

150 "Centre for the Fourth Industrial Revolution." World Economic Forum. https://www.weforum.org/centre-for-the-fourth-industrial-revolution. Accessed 27 Sep 2021.

151 "How Will Your Factory Transformation Strategy Reinvent Your Future?" EY. https://www.ey.com/en_us/advanced-manufacturing/how-will-your-factory-transformation-strategy-reinvent-your-future?WT.mc_id=10814360&AA.tsrc=paidsearch&s_kwcid=AL!10073!3!500541928933!b!!g!!%2Bsmart%20%2Bmanufacturing. Accessed 9 Sep 2021.

152 "Fortune 500." Fortune. https://fortune.com/fortune500/2020/search/. Accessed 9 Sep 2021.

153 "The Manufacturers Powering Productivity and Sustainability." McKinsey & Company. https://www.mckinsey.com/featured-insights/themes/the-manufacturers-powering-pro-ductivity-and-sustainability?cid=other-eml-alt-mip-mck&hdpid=e210ef64-eb1d-4bff-bae0-ee3efba48f99&hctky=13060269&hlkid=a774e1a8976d4a078406df70a8406fd5. Accessed 27 Sep 2021.

154 "Road Map." Cambridge Dictionary. https://dictionary.cambridge.org/dictionary/english/road-map. Accessed 1 Oct 2021.

155 "Clean Energy Smart Manufacturing Innovation Institute Roadmap 2017–2018." CESMII—The Smart Manufacturing Institute. https://www.cesmii.org/wp-content/uploads/2021/03/cesmii_roadmap_8–28–2017_pdf.pdf. Accessed 1 Oct 2021.

156 "Magic Quadrant for Robotic Process Automation." Gartner. https://www.gartner.com/account/signin?method=initialize&TARGET=http%3A%2F%2Fwww.gart-ner.com%2Fdocument%2F4016876%3Ftoggle%3D1%26refval%3D339540878%26ref%3DsolrAll. Accessed 18 Sep 2022.

157 "Gartner Identifies Key Emerging Technologies Spurring Innovation Through Trust, Growth and Change." Gartner. https://www.gartner.com/en/newsroom/press-releases/2021–08–23-gartner-identifies-key-emerging-technologies-spurring-innovation-through-trust-growth-and-change#:~:text=The%20Hype%20Cycle%20for%20Emerging%20Technologies%2C%202021%20is,Cycles%3A%20Innovating%20Delivery%20Through%20Trust%2C%20Growth%20and%20Change.%E2%80%9D. Accessed 21 Sep 2021.

158 "Kübler-Ross Change Curve®." The Elisabeth Kübler-Ross Foundation. https://www.ekrfoundation.org/5-stages-of-grief/change-curve/. Accessed 19 Oct 2021.

159 "Hype Cycle for Emerging Technologies, 2022." Gartner. https://www.gartner.com/ account/signin?method=initialize&TARGET=http%3A%2F%2Fwww.gartner.com %2Fdocument%2F4016848%3Fref%3DsolrAll%26refval%3D339541230%26toggl e%3D1. Accessed 18 Sep 2022.

160 "Closing the Employability Skills Gap." Deloitte. https://www2.deloitte.com/us/ en/insights/focus/technology-and-the-future-of-work/closing-the-employability-skills-gap.html. Accessed 1 Oct 2021.

161 "Our Mission and Work." CASEL. https://casel.org/about-us/our-mission-work/. Accessed 18 Sep 2022.

162 "About Explore SEL." Harvard University. Explore SEL. http://exploresel.gse. harvard.edu/about/. Accessed 1 Oct 2021.

3 Transformational Technologies

Adoption and Integration in Operating Plants

Hebab A. Quazi

3.1 INTRODUCTION

Smart Manufacturing technologies and techniques are necessary for the existing industries for their profitability, competitiveness, and sustainability planning. It is also required for product re-alignment, market restructuring, and adoption and integration of transformative technologies. The adoption and integration of transformative technologies utilizes appropriate industry data intelligently. Digitized processes are the foundation of smart manufacturing. The benefit of digital transformation is known for quite some time. The costs, schedule to implement, and ROI are the challenges. Some manufacturing industries have taken cautious steps and ventured into adopting smart manufacturing practices. The industries sell their products in competitive markets at a price paid by the customers. Satisfied customers buy the products repeatedly generating revenues with profits, year after year. The profits must be on a long-term basis to provide sustainability.

3.1.1 STRATEGIC PLANNING

Transformational Technologies need to survive the strategic planning phase that involves market restructuring as required, sustainability-roadmap planning, risks mitigation, intellectual-property protection, regulatory compliance, fundraising, and product re-launch planning. Applications of Industry-Best-Practices include Program Development and Implementation, Planning, Budgeting and Budget Control, Quality Assurance and Quality Control, Scheduling and Schedule Control, Cost Estimating and Cost Control, Testing and Validation, as required.

The industries consolidate their relevant business and plant data and engage IT and OT analytical specialists to develop their roadmap(s) to plan success. These roadmap(s) do provide appropriate direction(s) for their future business activities. If these activities are not done properly, it can lead to wasteful efforts. Sometimes, the manufacturers engage specialty firms for bringing in external expertise and know-how. Transformative manufacturing technologies can provide competitive advantages and sustainability. Each industry has its own strategy in owning and protecting its technological assets and data. These industries navigate their

DOI: 10.1201/9781003156970-3

pathways in managing their proprietary data and/or developing further technological directions for their own business success. Sometimes, intra-company exchange brings short-term and/or long-term benefits. Also, information-sharing with industry experts benefit developing cybersecurity and regulatory compliance strategy.

Many industries are still resisting adopting digitization. Sometimes, the information on ROI for certain industries is not easily available. Industries do have to develop their own ROI forecasts for their digital-transformation projects; otherwise, they may have difficulties in securing management approval to proceed. It is suggested that each company prepares its own "Strategic Plan" for Transformational Technology adoption for profitability, competitiveness, and sustainability. This requires a focused and dedicated team with a leader having adequate and appropriate experience. If and when required, the manufacturing company can engage an outside resource company for assistance or guidance.

Recent experience has shown that Industry 4.0 is a good roadmap for growth and profitability. A recent industry forecast shows that Industry 4.0 roadmap would give about 15% annual growth rate from 2020 to 2027. However, the growth rate for Transformational Technology adoption can be different and challenging.

3.1.2 PLANNING TEAM

It is prudent to select a competent team for the purpose. Also, consider adding external expertise or resources for the planning and execution of the program. Digital transformation can and (most likely) will make significant impact on the operation of the plant. The digitization process should make the industrial operation more efficient, profitable, competitive, and sustainable. To make these happen, the plant needs concerted efforts of the Board, Management, Operation and Supply-Chain leadership. The Planning team needs to appraise the ROI of the operation periodically. The question of any replacements (hardware/software) with costs and timing would have to be addressed in the plan. Industry management support system for the proposed changeover is critical to the digitization program.

Adoption and integration of transformational technologies in the operating plant begin with selection of appropriate techniques in developing the strategy, plans, and programs. Most often, the industry borrows money for this phase. Private and commercial lenders require that the technology provider convinces that the industry is capable of repaying. Other major step includes selecting appropriate manufacturing processes to increase the viability of the venture. Once the technological steps are established, it is very important to protect the intellectual property rights through patenting.

3.1.3 BUSINESS PLAN

The Business Plan must include at least the problem definition and identification, solution (product/services) proposed, target market(s) with customers, sales/marketing strategy, business model, identified competitions, competitive advantages, financial projection, resources (technical, management, marketing, financial, and regulatory), and target schedule. The planning section tells us what needs to be done and when and what benefits it will bring for satisfying the overall objectives.

The planning document is sometimes called the Project Execution Plan (PEP). This planning document normally consists of (1) Definition of the Tasks to be performed, (2) Statements on the Objectives of the Venture, (3) Identification of Potential Problems, (4) Lists of Assumptions, (5) Definitions of the Objectives, (6) Statements on Strategies, (7) Identifying Sequence of the Tasks, (8) Establishing Required Resources, (9) Reviewing and Finalizing the Plan, (10) Establishing Probability of Success, (11) Finalizing the Plan, after a Detailed Review, and (12) Reviewing of the Plan Often to Improve, as background data do change. It will be a prudent approach to review the Plan at least every 6 months and adjust the Plan where appropriate. Situations (particularly market and financing) controlling commercial ventures do change frequently.

3.1.4 FINANCIAL PLANNING

To attract investors, it is very important to prepare a believable plan including sales/revenue forecasts. Quite often, the owner of the venture is not equipped to develop a successful financial plan. It may be prudent to retain a Financial Advisor to work with the owner (or the company management). The Financial Advisors are knowledgeable about investor's preferences in all segments of private and public capital markets. The feasibility of financing will be based on answering certain questions. These are:

1. What are each shareholder's reasons for participating in raising capital?
2. What share of ownership does each existing shareholder wish to retain?
3. What level of risk each existing shareholder will accept?
4. Is the cash flow forecasted sufficiently justifying each shareholder's investment?
5. Will the anticipated cash flow be sufficient to attract capital from outside?

The appropriate tax structure(s) in a country will have significant impact on attractiveness for the investment.

The success of an emerging technology adoption is to be monitored, adjusted, secured, and assured often. Critical components in technology adoption roadmap includes as a minimum, technology development, financing, manufacturing, regulatory compliance, and customer acceptance. The management team must have adequate knowledge and experience to adjust the action plan as the situation changes.

3.1.5 VALUE PROPOSITION

Industry's current operational situations should be defined first in detail so that the cost associated with each step can be estimated. This is essential so that capital expenditure and their justifications can be prepared for investors or management consideration and approval. The proposed solution also should be broken down into details so that it is easy for the approval authorities to understand the measurable benefits. The challenges industry face include establishing convincing value propositions if any prototyping steps are required and then developing the financial justification in a business case for acceptance by the investor or by the company top management. The business case must include short-term and long-term benefits and risks mitigation or avoidance decision(s). The justification should include at least situational analysis for financial requirement.

The capital-expenditure proposal identifies clearly the value-added benefits including net profits that can be achieved and the time that will take before the benefits can be realized. Alternatives considered, if any, must be clearly stated along with their associated costs and potential benefits and risks. It is very important to demonstrate that fair and sound evaluation can be conducted. All potential solutions, particularly requiring capital-expenditures, have associated risks. The risks must be assessed clearly, and the degree of risks ranked. Risks analysis need to have, where appropriate, recommendations for risk aversion or mitigation.

3.1.6 RETURN-ON-INVESTMENT

The business plan's core objective is the ROI estimate on the capital requested. The ROI justification normally has two components: (1) Capital investment requirements including expenses and (2) Financial benefits including ROI or pay-back-period. The best-case scenario would show large return with small investment. The top strategy would be to select the one with most compelling benefits with good return that requires minimum investment. Consider the measurability of the proposed ROI as the first pick. To increase the probability of success in securing the approval of funding, identify the hardware that can be reused in the operation. Transformational technologies that have potentials include application of artificial intelligence (AI), machine learning, and Industrial Internet of Things (IIoT). The industry guidelines provide consistent management of manufacturing facility operational hazards and risks to the plant personnel. Sometimes, each manufacturing company develops its own guidelines and practices. These guidelines are practiced in all of their facilities for consistency. The Program Implementation Schedule is set by the industry.

3.1.7 CUSTOMER CARE

Business success is created through the eyes of the customers. Proper customer care helps creating better revenue and profit performance. Customer experience can influence the perception of the product and viability of the organization. Many organizations have developed several techniques on how to improve the customer perception of the product, its pricing, competitiveness, and sustainability. A good strategy is to collect customer feedback, understand the issues, and create a program that will not only maintain the level of confidence of the customers but also improve on it gradually or stepwise. The program must address both tactical and strategic issues.

3.1.8 ROADMAP PLANNING

Manufacturing roadmaps should result in much higher annual growth with more focused competitiveness and sustainability. Securing large profitability with better competitiveness and extended sustainability would require prudent integration of business acumen and experience. Accelerating the operational efficiency in an industry is a vital step in the digitization process. Small steps like adding new sensors in appropriate locations could be a prudent approach. Accuracy and validation of collected data can be another small step to success. Adding a historian to the plant having DCS or PLCs would be a good approach. Also, considerations should be given on how to automate a particular section of the plant that requires digitization. In the planning for digitization, the future operational plan needs to be spelled out. The industry expertise should be included in developing the digitization plan. Finally, the costs for digitization should be identified and ROI be calculated.

3.2 TECHNOLOGY ADOPTION AND INTEGRATION

Technology adoption and integration require careful consideration. If there is no due consideration, it can lead to success or failure. The planning and execution processes are rigorous and should be taken care of properly and in detail.

3.2.1 ADOPTION CONSIDERATIONS

The transformational technology adoption in a business or in a venture needs to go through a stepwise process including:

1. Identifying and/or defining the issue or problem.
2. Developing or creating a solution.
3. Describing competitive advantages of the proposed solution.
4. Selecting a technology or technologies for the proposed solution.
5. Assessing its current technology readiness level.
6. Developing commercialization plan, if required.
7. Arranging appropriate financing option(s).
8. Building prototypes for testing, if required.
9. Manufacturing or building hardware or system.
10. Testing market and collecting customer feedback.

3.2.2 INTEGRATION CHALLENGES

When considering integration or insertion of transformational technology into an existing industry, then appropriate emerging technologies such as smart sensors, AI, advanced analytics, and IIOT with cyber security should also be considered. For technology insertion in an existing manufacturing plant, considerations must be given on industry priorities. These priorities including improving plant reliability, operability, competitiveness, and profitability should also be checked. The industries should also eliminate interruption and wastes, where possible.

The manufacturing plant needs to adopt agile methodologies and techniques. The innovative technologies such as AI, Machine Learning, and IIoT are helping manufacturing plants become more efficient and competitive, especially in the oil and gas, chemical, petrochemicals, mining, and manufacturing sectors.

3.3 PERFORMANCE CHALLENGES

Performance monitoring and forecasting are considered challenging in running a business. So, the industry must pay attention to Economics and Pricing, Technological

Opportunities, and Customer Preferences. The industry must monitor or track Competitors' Programs and Regulatory Challenges.

3.3.1 ECONOMICS AND PRICING

To customers, pricing of a product is very important. Pricing involves affordability and competitiveness. For any venture, it is essential to understand why customers buy the product at a price they are willing to pay. One of the many challenges in understanding your own performance is to evaluate and compare competition's promotional programs.

Based on market intelligence, the timing and content of your own promotional programs should be developed and implemented. It is very important to understand the impacts of your own promotional activities. Collecting and understanding customers' feedback is critical as immediate actions can be taken for any negative comments. So, it is important to develop a system for engaging the customers regularly. Also, customers can tell us the negative aspects of your competitors' products.

3.3.2 TECHNOLOGICAL OPPORTUNITIES

AI is being applied in the marketplace to secure greater benefits. AI in association with machine learning is now a powerful tool for the industries. AI along with IIoT, helped develop "Industry 4.0" platform. These innovative technologies are helping manufacturing plants become more efficient and competitive.

3.3.3 CUSTOMERS' PREFERENCES

Customers are key to a business or a venture. It is very important to understand why the customers buy the products, and more importantly the answers to "why" questions. Customers' loyalty is key, but the most important information is why they change their preferences, when it happens. Most often it is the price, but sometime the competitor's other aspects of their offerings become more attractive, may be it is cultural.

3.3.4 COMPETITORS' PROGRAM DYNAMICS

We need to recognize that our competitors are watching us all the time. We must watch and predict all possible reasons why, when, and how often our competitors change their marketing plans. It is utmost important to prepare our own program to win over our competitors.

3.3.5 REGULATORY CHALLENGES

We need to observe and understand how (why, and how often) certain country's/state's regulations change. Usually, once certain regulations are enacted, they are expected to continue for a while. Sometimes, it is possible to forecast certain changes in regulations and their timing. Most often, the trade-barriers, environmental considerations, and product compositions are the reasons for the regulations and their changes.

3.4 COMPETITIVENESS TRACKING

Business success is created through the eyes of customers. Proper customer care helps to develop realistic revenue forecast and profit performance. Customer experience can influence the perception of the product affecting the viability of the organization. Many organizations have developed a number of techniques on how to improve the customer perception on products, its pricing, competitiveness, and sustainability. Good strategy must include collecting customer feedback, understanding of their issues, and creating programs that not only will maintain the level of confidence of the customers but also improve it gradually or stepwise. The program must address both tactical and strategic issues.

3.4.1 TRACKING COMPETITIONS

For customers, pricing of the product is very important. This involves affordability and competitiveness. One of many challenges is to evaluate the pricing of the competitions and keeping a lookout for their promotional programs. Based on market intelligence, the timing and content of its own promotional programs should be developed and implemented. Also, it is important to understand the impact of own promotional activities. For any venture, it is essential to understand why a customer buys its product at a price they are willing to pay. It is very important to get customers' feedback, so that immediate actions can be taken for any negative comments. It is essential to develop a system that engages the customers regularly. Customers can also tell us the negative aspects of our competitors' products.

3.4.2 MARKET TRENDS

The technology adoption challenges need to be checked against the market trends. Market risks must be assessed regularly and appropriate recommendations for risk aversion or mitigation are essential for an existing operating venture. Transformational technology adoption in any commercial ventures can include certain intermediate steps including assessing technology readiness, risks mitigation, sustainability planning, manufacturing, financing, marketing, and customer acceptance.

3.4.3 GLOBAL ECONOMIC FORECASTS

Global economic forecasts can sometimes be quite complex with regional interdependencies. Certain products can have greater regional interdependencies including seasonal raw materials availability, regional economies, production cycles, and customers' preferences. It is recommended that regular attention must be given to monitor and understand the regional and global economic forecasts.

3.5 SUSTAINABILITY CHALLENGES

The Sustainability Challenges should include at least the following issues: (1) Technology development consideration, if applicable, (2) Testing and manufacturing resources, (3) Financing support (as/when needed), (4) Market intelligence, (5) Management continuity, (6) Third-party resources for hire, and (7) Industry's economic conditions. The Sustainability Challenges could include a "Project-Lifecycle Roadmap." The Lifecycle Roadmap development needs expertise that is scarce in most ventures, especially for the first-time adoption cases.

3.5.1 MARKET FORECASTS

The market forecasts that are essential for creating sustainable values should include entrepreneurial leadership in building commercial ventures, teaming for sustainability, and customer care essentials. The key market forecasting expertise is sometimes difficult to have in-house for a specific product segment or launching a product for a very special customer segment. Also, inability to understand the market forecast for any specific customer segment can cause business venture failures. Depending on the market segment, a marketing specialist with appropriate experience should be engaged. The experts are not available all the time, especially when you need them.

3.5.2 COMPETITIVENESS TRENDS

Competitiveness trends are established by the industry leaders and individual operating plants try to follow it. Each operating plant needs to decide whether to become the leader by establishing the trends or be the follower.

3.5.3 CUSTOMERS' PREFERENCES

Customers help establish the trends through their preferences. Most of the customer preferences are predictable, but certain customers preferences are difficult to predict. Customer care programs are developed based on their preferences. As the customer preferences tend to change, the marketing plans need to be adjusted accordingly.

3.5.4 PERFORMANCE TRENDS

Industries set their group trends in performances. Each industry or operating plant tries to compare their performances against the industry trends. Certain industry leads the trend-in-performance, or even target to exceed industry performance.

3.6   SMART MANUFACTURING ROADMAPS

The primary steps in preparing the industry roadmaps require careful planning. It is very important to understand the inter-relationship between the roadmaps. All the roadmap segments constitute the total roadmap for the venture.

3.6.1 STRATEGIC OPTIONS

   (a) Innovation and Improvisation
   (b) Improve Performance and Agility

(c) Assure Transformational Technologies Adoption

(d) Facilitating Integration of Technologies in Operational Network

(e) Improving Quality and Mitigating Risks

(f) Future-Proofing Customers' Acceptance

3.6.2   OPERATIONAL CHANGES

3.6.2.1   Operational Modifications Requirements

(a) Restructure Operations to Improve

(b) Processing Improvement and Betterment

(c) Re-Specifying Raw Materials, if needed

(d) Improve Supply-Chain Challenges

(e) Quality Improvement

(f) Costs Reduction

(g) Wastes Reduction

3.6.3   COMPETITIVENESS CHALLENGES

(a) Technology and business strategic

(b) Re-evaluating market (size, trends, and sustainability)

(c) Identify and evaluate competitions

(d) Re-establishing competitive products, market shares, pricing trend

(e) Re-evaluate customer preferences and pricing structure

(f) Re-visit strategies for new market entry

(g) Identifying appropriate market entry methodologies

(h) Re-evaluate own-company strengths and weaknesses

(i) Identify market (local/regional/cultural) preferences if any

(j) Re-establish economy of scales, logistic challenges, and regulatory challenges.

3.6.4   SUSTAINABILITY CHALLENGES

(a) Technology and business strategy

(b) Feasibility verified (short term and long term)

(c) Revenue optimized (short term and long term)

(d) Check on external resources' availability

(e) Financing assured or availability

(f) Constructability challenges, if any

(g) Reestablish customer feedback (short term and long term)

(h) Provide or obtain regulatory trends

3.6.5   QUALITY, SAFETY, AND SECURITY

(a) Quality assurance

(b) Prioritizing safety

(c) Mitigating security risks

3.6.6 PERFORMANCE FORECASTING

(a) Quality
(b) Safety
(c) Risks
(d) Pricing
(e) Profitability
(f) Competitiveness
(g) Sustainability

4 Roadmaps for the Adoption and Integration of Transformational Technologies in Key Manufacturing Plants

Hebab A. Quazi

4.1 INTRODUCTION

This chapter demonstrates typical roadmaps for the adoption and integration of transformational technologies in key industrial plants. Three different types of industrial plants are picked to demonstrate complexities in manufacturing. The complexities vary widely for different parts of the world. These manufacturing industries include Petroleum Refining, Plastics Products Manufacturing, and Pulp and Paper Industry. The readers are advised to use their judgments in using these roadmaps.

Petroleum Fuels are widely used in different parts of the world and the refineries are designed and built to meet the local and regional needs. Crude oil is produced in different parts of the world having deferent compositions and impurities. Crude oils from different sources are mixed/blended before refining. Petroleum fuels used in different parts of the world have different specifications and are governed by countries' own regulations. Emissions from petroleum fuel combustions can vary, sometimes widely. Each country has environmental emissions control regulations that must be met. These regulations control global warming emissions such as methane and carbon dioxide.

Plastic Product Manufacturing plants are normally established downstream of petrochemical plants. Raw materials for plastic product plants such as polyethylene and polystyrene, polypropylene, PVC, and other polymers come from the petrochemical plants. Marketing of plastic products are quite competitive and are transported by land transportation. Plastic product manufacturing plants need to operate within the country's local regulations including for emissions control and waste management. Plastic product manufacturing is growing fast and offering varieties of products, utilized by other industries as well. Plastic product industries are designed and built to meet local and regional consumers' needs and to meet local environmental regulations.

Pulp and Paper Plants are normally built close to the sources of their raw materials, most likely near the woods or forests. Their products are heavier for transportation and distribution. Marketing and consumption are limited to regional needs. These plants usually are polluting and regional regulations are created to curb the pollution.

Suggestions for readers include development of their own plant roadmaps after understanding of their industry roadmaps. The roadmaps presented here are created as examples. Also, the activities in each Roadmap are presented as suggestions. Readers are encouraged to take these as examples and use for their own plant application.

4.2 PETROLEUM REFINING ROADMAPS

4.2.1 Petroleum Refining Roadmap – 1

4.2.1.1 Re-Visiting Goals and Objectives
Target Competitive Advantages with Transformational Technologies.
Maximize Potential Benefits in Crude Oil Procurement.
Establish Crude Oils Blending Range for Profitability, Market Share, and Regulatory Compliance.
Improve Performance and Operational Agility of Refining Operations.
Create Options with Crude Mix, Processing Capability, Product Quality.
Target Transformational Technologies Adoption and Integration Capabilities.
Maximize Operational Flexibility.
Assure Customers Comforts on Pricing, Availability, Quality, and Competitiveness.
Mitigating Risks on All Areas.

4.2.2 Petroleum Refining Roadmap – 2

4.2.2.1 Operational-Modifications Planning
Restructure Operating Parameters for Maximizing Benefits.
Ensure Transformational Technologies' Adoption and Integration.
Optimize Re-boilers Operations to Minimize Steam/Energy Uses.
Improve Condensers' Performance for Product Quality.
Automate Operations of Distillation Columns and Condensers for Performance Improvement.
Improve Predictive Performances of Target Equipment and their Maintenance Operations.
Maximize Use of Advanced Sensors, AI, and Machine Learning Technologies.
Adopt Digital Twin Technology Concept, where Possible.
Overcome Supply-Chain Difficulties.
Improve Energy Efficiencies, Possibly Generate Power with Excess Energy, as a Bi-Product.
Minimize Waste Generation.

4.2.3 Petroleum Refining Roadmap – 3

4.2.3.1 Competitiveness Challenges
Identify Transformational Technology Application(s).

Maximize Use of Advanced Sensors, AI, and Machine Learning Technologies.
Adopt Digital Twin Technology Application, where Possible.
Position Appropriately in Strategic Markets (Size, Trend, and Sustainability), Locally and Globally.
Identify and Evaluate Current and Future Competitions in Petroleum Products.
Re-Structure Marketing Strategy for Competitive Products, Market Shares, and Pricing Trends.
Evaluate Customer Preferences and Pricing Structure.
Establish Economy of Operations for Multiple Products.
Resolve Local and Regional Transportation Challenges.
Forecast Local and Regional Regulations, where Possible.
Develop Strategies for New Product Market Entry (Local as well as Global) for Refined Products.

4.2.4 PETROLEUM REFINING ROADMAP – 4

4.2.4.1 Sustainability Challenges

Identify Roadmaps for Refined Products for the Existing and Potential Markets.
Track Transformational Technologies for Current and New Markets for Green(er) Products.
Develop Technologies for Converting Waste Petroleum Stream(s) to Marketable Products.
Verify Feasibility for New (and Green) Products.
Establish Roadmaps for Profitability for All Refined Products.
Identify Internal as well as External Sources for Financing.
Secure Skilled Manpower to Optimize Operation and Assets Utilization.
Engage External Experts, if Required, for Competitive Energy Transition.
Reestablish Customer Feedback Systems for New and Existing Product Lines.
Forecast Regulatory Trends, for the Existing and New Refined Products.

4.2.5 PETROLEUM REFINING ROADMAP – 5

4.2.5.1 Quality, Safety, and Security Management

Assure Product Quality Trends.
Prioritize Personnel Safety.
Mitigate Safety Risks.
Verify IT/OT/IIoT and Digitization.
Eliminate or Mitigate Operating Systems Risks.
Identify and Implement Cybersecurity Control Tools.

4.2.6 PETROLEUM REFINING ROADMAP – 6

4.2.6.1 Performance Targets

Production
    Next Year.
    Production in Each Year through the Fifth Year.

Quality
   Next Year.
   Five Years from Now.
Safety and Security
   All through the Next 5 Years.
Risks
   Risks Mitigated All through to the Next 5 Years.
Pricing
   Pricing of All Products through to the Next 5 Years.
Profitability
   Estimate Profits in Each of the Next 5 Years.
Competitiveness
   Achieve No! Position in xx Years.
Sustainability
   Stay Profitable in Each of the Next 5 Years.

4.3 PLASTIC PRODUCT MANUFACTURING ROADMAPS

4.3.1 PLASTIC PRODUCT MANUFACTURING ROADMAP – 1

4.3.1.1 Re-Visiting Goals and Objectives

Target Competitive Advantages with Transformational Technologies.
Maximize Potential Benefits in Each Product Line or Business.
Plan for Plastic Product Lines' Profitability, Market Share, and Regulatory Compliance.
Improve Performance and Operational Agility of Manufacturing Operations.
Create Options with Plastic Product Mix, Manufacturing Capability, and Product Quality.
Target Transformational Technologies Adoption and Integration Capabilities.
Maximize Operational Flexibility.
Assure Customers Comforts on Pricing, Availability, Quality, and Competitiveness.
Mitigating Risks for All Product Lines and All Markets.

4.3.2 PLASTIC PRODUCT MANUFACTURING ROADMAP – 2

4.3.2.1 Operational-Modifications Planning

Restructure Operating Parameters for Maximizing Benefits.
Ensure Transformational Technologies' Adoption and Integration.
Optimize Equipment Operations to Minimize Energy Uses.
Improve Product Quality.
Automate Operations for Performance Improvement.
Improve Predictive Performances of Target Equipment and their Maintenance Operations.
Maximize Use of Advanced Sensors, AI, and Machine Learning Technologies.

Adopt Digital Twin Technology Concept, where Possible.
Overcome Supply-Chain Challenges.
Improve Raw Materials Utilization.
Minimize Waste Generation.
Check Possibility of Bi-Product(s) Manufacturing and Marketing.

4.3.3 PLASTIC PRODUCT MANUFACTURING ROADMAP – 3

4.3.3.1 Competitiveness Challenges

Identify Targets for Transformational Technology Application(s).
Maximize Use of Advanced Sensors, AI, and Machine Learning Technologies.
Adopt Digital Twin Technology Application, where Possible.
Position in Strategic Markets (Size, Trend, and Sustainability).
Identify and Evaluate Current and Future Competitions in All Plastic Product Lines.
Re-Structure Competitive Products, Market Shares, and Pricing Trends.
Evaluate Customer Preferences and Pricing Structure.
Establish Economy of Operations for Different Products.
Evaluating and Resolving Local and Regional Plastic Product Transportation Challenges.
Forecasting Local and Regional Regulations, where Possible.
Develop Strategies for New Market Entry (Local, Regional, as well as Global), as Applicable.

4.3.4 PLASTICS PRODUCTS MANUFACTURING ROADMAP – 4

4.3.4.1 Sustainability Challenges

Identify Roadmap for All Plastic Product Manufacturing.
Track Transformational Technologies that would Help Entering New Markets.
Restructure Business Operations Required for the Next 5 Years.
Verify Restructured Operations' Feasibility.
Establish Roadmap for Profitability.
Work with External Expert Resources.
Identify External Sources for Financing Restructured Operation, If Needed.
Overcome any Constructability Challenges in Plant Restructuring.
Secure Appropriate Skilled Manpower, if Required (Secure Expertise from External Sources).
Reestablish Customer Feedback Systems.
Forecast Regulatory Trends.

4.3.5 PLASTIC PRODUCT MANUFACTURING ROADMAP – 5

4.3.5.1 Quality, Safety, and Security Management

Assure Product Quality Trends.
Prioritize Personnel Safety.

Mitigate Safety Risks.
Verify IT/OT/IIoT and Digitization Options.
Mitigate Operating Systems' Risks.
Identify and Implement Cybersecurity Tools.

4.3.6 PLASTIC PRODUCT MANUFACTURING ROADMAP – 6

4.3.6.1 Performance Targets

Production
   Next Year.
   Production in Each Year through the Fifth Year.
Quality
   Next Year.
   Five Years from Now.
Safety and Security
   All through the Next 5 Years.
Risks
   Risks Mitigated All through to the Next 5 Years.
Pricing
   Pricing of All Products through to the Next 5 Years.
Profitability
   Profits in Each of the Next 5 Years.
Competitiveness
   Achieve No 1 Position in xx Years.
Sustainability
   Stay Profitable in Each of the Next 5 Years.

4.4 PULP AND PAPER INDUSTRY

4.4.1 PULP AND PAPER INDUSTRY ROADMAP – 1

4.4.1.1 Re-Visiting Goals and Objectives

Re-Visit Availability of Local and Regional Fibers for Pulping and Papers.
Create Options with Applicable Fiber Mix, Processing Capability, and Product
   Quality.
Maximize Potential Benefits with Recycling Residuals.
Establish Fiber-Blended Products for Profitability, Market Share and Regulat-
   ory Compliance.
Improve Performance and Operational Agility of the Industry Operations.
Target Transformational Technologies' Adoption and Integration Capabilities.
Maximize Operational Flexibility.
Assure Customer Comforts on Pricing, Availability, Quality, and Competiti-
   veness.
Mitigating Risks on All Areas.

4.4.2 PULP AND PAPER INDUSTRY ROADMAP – 2

4.4.2.1 Operational Modifications Planning

Restructure Operating Parameters for Maximizing Benefits.

Ensure Transformational Technologies' Adoption and Integration.

Optimize Digesters' and Driers' Operations to Minimize Steam/Energy Uses.

Improve Industry Performance for Products Quality.

Automate Operations for Performance Improvement.

Improve Predictive Performances of Target Equipment and their Maintenance Operations.

Maximize Use of Advanced Sensors, AI, and Machine Learning Technologies.

Adopt Digital Twin Technology Concept, where Possible.

Overcome Supply-Chain Difficulties.

Improve Energy Efficiencies.

Minimize Waste Generation.

4.4.3 PULP AND PAPER INDUSTRY ROADMAP – 3

4.4.3.1 Competitiveness Challenges

Identify Transformational Technology Application(s).

Maximize Use of Advanced Sensors, AI, and Machine Learning Technologies.

Adopt Digital Twin Technology Application, where Possible.

Position the Industry in Strategic Markets (Trend and Sustainability), Locally and Regionally.

Identify and Evaluate Current and Future Competitions in Paper Products.

Re-Structure Marketing Strategy for Competitive Products, Market Shares, and Pricing Trends.

Evaluate Customer Preferences and Pricing Structures.

Establish Economy of Operations for Multiple Products.

Evaluate Local and Regional Transportation Challenges.

Forecast Local and Regional Regulations, where Possible.

Develop Strategies for New Product Market Entry (Local as well as Regional).

4.4.4 PULP AND PAPER INDUSTRY ROADMAP – 4

4.4.4.1 Sustainability Challenges

Identify Roadmaps for Paper Products from Existing and Potential New Raw Materials.

Track Transformational Technology Application for the Current and New Paper Products.

Develop Technologies for Converting Waste to New Paper Products.

Verify Feasibility for New (and Greener) Products.

Establish Roadmap for Profitability for All Paper Products.

Identify Internal as well as External Sources for Financing.

Secure Skilled Manpower to Optimize Operation and Assets Utilization.
Engage External Experts, if Required, for Entering New Markets.
Reestablish Customer Feedback Systems for New and Existing Product Lines.
Forecast Regulatory Trends for the Existing and New Products.

4.4.5 PULP AND PAPER INDUSTRY ROADMAP – 5

4.4.5.1 Quality, Safety, and Security Management
Assure Product Quality Trends.
Prioritize Personnel Safety.
Mitigate Safety Risks.
Verify IT/OT/IIoT and Digitization.
Mitigate Operating Systems' Risks.
Identify and Implement Cybersecurity Control Tools.

4.4.6 PULP AND PAPER INDUSTRY ROADMAP – 6

4.4.6.1 Performance Targets
Production
   Next Year.
   Production in Each Year through the Fifth Year.
Quality
   Next Year.
   Five Years from Now.
Safety and Security
   All through the Next 5 Years.
Risks
   Risks Mitigated All through to the Next 5 Years.
Pricing
   Pricing of All Products through to the Next 5 Years.
Profitability
   Profits in Each of the Next 5 Years.
Competitiveness
   Achieve No. 1 Position in xx Years.
Sustainability
   Stay Profitable in Each of the Next 5 Years.

Appendix 1

The following information is provided in support of statements made herein and is provided for ease of readership. Statements and definitions herein are attributed to sources cited and are the opinion and position of those organizations.

GLOSSARY

The following terms are used in this report and by practitioners. These definitions are provided for the convenience of the readers.

Note: As a general rule, the following Glossary term definitions are taken from the relevant websites and are references as appropriate. As used throughout this report, direct quoted citations are in *italics*, per editorial convention.

GLOSSARY

Term	Definition
Acculturation	*Cultural modification of an individual, group, or people by adapting to or borrowing traits from another culture.[1]*
Actionable	*Items are tasks that are defined for issues that facilitate issue resolution.[2]*
Artificial Intelligence (AI)	*The capability of a machine to imitate intelligent human behavior.[3]*
Behavioral Economics	*An approach to economic analysis that incorporates psychological insights into individual behavior to explain economic decisions.[4]*
Blockchain	*A growing list of records, called blocks, that are linked together using cryptography.[5]*
Bridging Document	*A written document which defines how two or more Management Systems co-exist to allow cooperation and coordination on matters of health, safety, and environmental protection between different parties, usually between an Authorized Person and its Sub-Contractors, cross-referencing the detailed procedures which will be used and defining the responsibilities, accountabilities, and work activities of the various parties.[6]*
Business Process Model	*A business process model is a model of one or more business processes and defines the ways in which operations are carried out to accomplish the intended objectives of an organization. It can describe the workflow or the integration between business processes. It can be constructed in multiple levels.[7]*
Capital Efficiency	*The continual improvement in the production output of a piece of capital equipment per unit of time.[8]*
Commercial Off the Shelf (COTS)	*Software and hardware that already exists and is available from commercial sources. It is also referred to as off-the-shelf.[9]*

(Continued)

Term	Definition
Consensus (settled) Science	*There is no such thing as consensus science. If it's consensus, it isn't science. If it's science, it isn't consensus. Period.*[10]
Continuous Improvement	*Sometimes called continual improvement, is the ongoing improvement of products, services, or processes through incremental and breakthrough improvements. These efforts can seek "incremental" improvement over time or "breakthrough" improvement all at once.*[11]
Creative Destruction	*Refers to the incessant product and process innovation mechanism by which new production units replace outdated ones. It was coined by Joseph Schumpeter (1942), who considered it "the essential fact about capitalism".*[12]
Critical Path	*The continuous string(s) of critical activities in the schedule between the Start and Finish of the project. The sum of the activity durations in the Critical Path is equal to the Project's Duration; therefore, a delay to any Critical Activity will result in a delay to the Project Completion Date.*[13]
Culture	*The set of shared attitudes, values, goals, and practices that characterizes an institution or organization.*[14]
Cybersecurity	*The practice of protecting systems, networks, and programs from digital attacks.*[15]
Data Bias	*Three types of bias can be distinguished: Information bias, selection bias, and confounding.*[16]
Data Reduction	*The transformation of numerical or alphabetical digital information derived empirically or experimentally into a corrected, ordered, and simplified form.*[17]
Defense in Depth	Commonly referred to from a military and cybersecurity perspective, it refers to multiple "layers" of defense.
Deterministic	*Of or relating to a process or model in which the output is determined solely by the input and initial conditions, thereby always returning the same results (opposed to stochastic).*[18]
Digitalization	*The process of converting something to digital form.*[19]
Economic Costs	*The accounting cost (explicit cost) plus the opportunity cost (implicit cost). Implicit cost refers to the monetary value of what a company foregoes because of a choice it made.*[20]
Economic Marginal Utility	*The additional satisfaction or benefit (utility) that a consumer derives from buying an additional unit of a commodity or service.*[21]
Economic Utility Theory	*Refers to the total satisfaction received from consuming a good or service.*[22]
Ecosystem	*The complex of living organisms, their physical environment, and all their interrelationships in a particular unit of space.*[23]
Empirical	*Relying on experience or observation alone often without due regard for system and theory.*[24]
Endogenous	*Caused by factors inside the organism or system.*[25]
Enterprise Resource Planning (ERP)	*A process used by companies to manage and integrate the important parts of their businesses. Many ERP software applications are important to companies because they help them implement resource planning by integrating all of the processes needed to run their companies with a single system.*[26]

(Continued)

Term	Definition
Exogenous	*Introduced from or produced outside the organism or system.[27]*
Expected Value (EV)	*The expected value (EV) is an anticipated value for an investment at some point in the future. In statistics and probability analysis, the expected value is calculated by multiplying each of the possible outcomes by the likelihood each outcome will occur and then summing all of those values. By calculating expected values, investors can choose the scenario most likely to give the desired outcome.[28]*
Expected Value of Marginal Information (EVMI)	*Expected value of the best decision with new information (obtained at no economic cost).[29]*
FBLearner Flow	*Facebook platform capable of easily reusing algorithms in different products, scaling to run thousands of simultaneous custom experiments, and managing experiments with ease.[30]*
Five (5) G	*Fifth generation technology standard for broadband cellular networks, which cellular phone companies began deploying worldwide in 2019.[31]*
Fourth Industrial Revolution (4IR)	*A fusion of advances in artificial intelligence (AI), robotics, the Internet of Things (IoT), genetic engineering, quantum computing, and more.[32]*
Future Proofing	*Design or change it so that it will continue to be useful or successful in the future if the situation changes.[33]*
Generally Accepted Practice (GAP)	*That either an authoritative accounting rule-making body has established a principle of reporting in a given area or that over time a given practice has been accepted because of universal application.[34]*
Harmonization (standards)	*The process of minimizing redundant or conflicting standards which may have evolved independently.[35]*
HSE	*Health, Safety, and Environment.[36]*
Human Factors	*The field which is involved in conducting research regarding human psychological, social, physical, and biological characteristics, maintaining the information obtained from that research, and working to apply that information with respect to the design, operation, or use of products or systems for optimizing human performance, health, safety, and/or habitability.[37]*
Innovation	*The introduction of something new.[38]*
Intellectual Property (IP)	*Creations of the mind, such as inventions; literary and artistic works; designs; and symbols, names, and images used in commerce.[39]*
Key Performance Indicators KPIs)	*Measure a company's success versus a set of targets, objectives, or industry peers.[40]*
Knowledge, Skills, and Abilities (KSAs)	*What is needed to do a job or task.*

(Continued)

Term	Definition
Learning Management System (LMS)	*A software application for the administration, documentation, tracking, reporting, automation, and delivery of educational courses, training programs, or learning and development programs.*[41]
Legacy Application or System	*An information system that may be based on outdated technologies but is critical to day-to-day operations.*[42]
Marginal Cost	*The change in the total cost that arises when the quantity produced is incremented, the cost of producing additional quantity.*[43]
Metaverse	*Alternate digital reality that would connect all the various proprietary digital spaces together.*[44]
Mindfulness	*The basic human ability to be fully present, aware of where we are and what we're doing, and not overly reactive or overwhelmed by what's going on around us.*[45]
Minimum Viable Product (MVP)	*A development technique in which a new product or website is developed with sufficient features to satisfy early adopters.*[46]
Mission Critical	*A task, service, or system is one whose failure or disruption would cause an entire operation or business to grind to a halt. It is a type of task, service, or system that is indispensable to continuing operations.*[47]
MITRE ATT&CK®	*A globally-accessible knowledge base of adversary tactics and techniques based on real-world observations. The ATT&CK knowledge base is used as a foundation for the development of specific threat models and methodologies in the private sector, in government, and in the cybersecurity product and service community.*[48]
Net Promoter Score (NPS)	*Widely used market research metric that typically takes the form of a single survey question asking respondents to rate the likelihood that they would recommend a company, product, or a service to a friend or colleague.*[49]
Object Management Group (OMG)	*An open membership, not-for-profit computer industry standards consortium that produces and maintains computer industry specifications for interoperable, portable, and reusable enterprise applications in distributed, heterogeneous environment.*[50]
One Throat to Choke	*An expression used in business to describe the advantage of purchasing goods or integrated services from a single vendor. That way, when something goes wrong, there is only "one throat to choke."*[51]
Open Source	*Software Source code that is made freely available for possible modification and redistribution. Products include permission to use the source code, design documents, or content of the product. The open-source model is a decentralized software development model that encourages open collaboration.*[52]
Operating Cost (OpEx)	*Includes both costs of goods sold (COGS) and other operating expenses – often called selling, general, and administrative (SG&A) expenses.*[53]
Operational Excellence	*The execution of the business strategy more consistently and reliably than the competition, with lower operational risk, lower operating costs, and increased revenues relative to its competitor.*[54]
Opportunity Cost	*The value of the next-highest-valued alternative use of that resource.*[55]

(Continued)

Term	Definition
Performance Management	*An ongoing process of communication between a supervisor and an employee that occurs throughout the year, in support of accomplishing the strategic objectives of the organization. The communication process includes clarifying expectations, setting objectives, identifying goals, providing feedback, and reviewing results.*[56]
PMBOK	*Project Management Body of Knowledge is a set of standard terminology and guidelines (a body of knowledge) for project management.*[57] Developed by the Project Management Institute (PMI).
Portfolio Management	*The selection, prioritization, and control of an organization's programs and projects, in line with its strategic objectives and capacity to deliver. The goal is to balance the implementation of change initiatives and the maintenance of business – as – usual, while optimizing return on investment.*[58]
Predictive Maintenance	*Techniques are designed to help determine the condition of in-service equipment in order to estimate when maintenance should be performed.*[59]
Proof of Concept (PoC)	*An effective proof of concept proves the goal of a proposed project is viable, and will be successful. The value of a POC is it can help a project manager identify gaps in processes that might interfere with success.*[60]
Reference Architecture	*A document or set of documents that provides recommended structures and integrations of IT products and services to form a solution. The reference architecture embodies accepted industry best practices, typically suggesting the optimal delivery method for specific technologies.*[61]
Reliability	*The extent to which an experiment, test, or measuring procedure yields the same results on repeated trials.*[62]
Research and Development (R&D)	*The part of a commercial company's activity concerned with applying the results of scientific research to develop new products and improve existing ones.*[63]
Return on Investment (ROI)	*A ratio that compares the gain or loss from an investment relative to its cost.*[64]
Right (License) to Operate	*Engaging in commerce in any form, including by acquiring, developing, maintaining, owning, selling, possessing, leasing, or operating equipment, facilities, personnel, products, services, personal property, real property, or any other apparatus of business or commerce.*[65]
Risk Mitigation	*The identification, evaluation, and prioritization of risks (defined in ISO 31000 as the effect of uncertainty on objectives) followed by coordinated and economical application of resources to minimize, monitor, and control the probability or impact of unfortunate events or to maximize the realization of opportunities.*[66]
Sarbanes-Oxley (SOX) Act of 2002	*Came in response to financial scandals in the early 2000s involving publicly traded companies such as Enron Corporation, Tyco International plc, and WorldCom. The act created strict new rules for accountants, auditors, and corporate officers and imposed more stringent recordkeeping requirements. The act also added new criminal penalties for violating securities laws.*[67]
SG&A	*Selling, general and administrative expense (SG&A) is reported on the income statement as the sum of all direct and indirect selling expenses and all general and administrative expenses (G&A) of a company. It includes all the costs not directly tied to making a product or performing a service.*[68]

(*Continued*)

Term	Definition
Single Sign-On (SSO)	*An authentication scheme that allows a user to log in with a single ID and password to any of several related, yet independent, software systems.[69]*
Situation Awareness	*Conscious knowledge of the immediate environment and the events that are occurring in it.[70]*
Smart Technologies	*The term "smart" is in fact an abbreviation for "Self-Monitoring Analysis and Reporting Technology." The technologies (include physical and logical applications in all formats) that are capable to adapt automatically and modify behavior to fit environment, sense things with technology sensors, thus providing data to analyze and infer from, drawing conclusions from rules. It also is capable of learning that is using experience to improve performance, anticipating, thinking, and reasoning about what to do next, with the ability to self-generate and self-sustain.[71]*
Standard(s)	*A set of specifications to which all elements of product, processes, formats, or procedures under its jurisdiction must conform.[72]*
STEM	*Science, Technology, Engineering, and Mathematics.[73]*
Stochastic	*Of or relating to a process involving a randomly determined sequence of observations each of which is considered as a sample of one element from a probability distribution.[74]*
Subject Matter Expert (SME)	*Provides the knowledge and expertise in a specific subject, business area, or technical area for a project.[75]*
Supervisory Control and Data Acquisition (SCADA)	*A control system architecture comprising computers, networked data communications and graphical user interfaces for high-level supervision of machines and processes. It also covers sensors and other devices, such as programmable logic controllers, which interface with process plant or machinery.[76]*
Supply Chain Management (SCM)	*The management of the flow of goods and services and includes all processes that transform raw materials into final products. It involves the active streamlining of a business's supply-side activities to maximize customer value and gain a competitive advantage in the marketplace.[77]*
Use Case	*A sequence of actions performed in a business that produces a result of observable value to an individual actor of the business.[78]*
Validity	*The quality of being well-grounded, sound, or correct.[79]*
Velocity of Information	*Similar to the economic theory, Velocity of Money, it is the frequency at which information is exchanged.[80]*
Velocity of Money	*A ratio of nominal GDP to a measure of the money supply (M1 or M2). It can be thought of as the rate of turnover in the money supply – that is, the number of times one dollar is used to purchase final goods and services included in GDP.[81]*
Wisdom of Crowds	*The idea that large groups of people are collectively smarter than individual experts when it comes to problem-solving, decision-making, innovating, and predicting.[82]*
Workflow Engine	*A software application that manages business processes.[83]*

For readers interested in more and greater detailed definitions of economics, see the Routledge Dictionary of Economics.[84]

NOTES

1 "Acculturation." Merriam-Webster.com Dictionary, Merriam-Webster, https://www.merriam-webster.com/dictionary/acculturation. Accessed 31 Aug 2021.

2 "Using Project Execution Management." Oracle. https://docs.oracle.com/en/cloud/saas/project-management/20b/oapem/project-issues.html#OAPEM813101. Accessed 23 Aug 2021.

3 "Artificial intelligence." Merriam-Webster.com Dictionary, Merriam-Webster, https://www.merriam-webster.com/dictionary/artificial%20intelligence. Accessed 1 Sep 2021.

4 "Overview: Behavioral economics. Oxford Reference. https://www.oxfordreference.com/view/10.1093/oi/authority.20110803095456532. Accessed 24 Sep 2021.

5 "Blockchain." Wikipedia. https://en.wikipedia.org/wiki/Blockchain. Accessed 23 Sep 2021.

6 "Bridging Document definition." Law Insider. https://www.lawinsider.com/dictionary/bridging-document. Accessed 4 Sep 2021.

7 "Business process modeling." Wikipedia. https://en.wikipedia.org/wiki/Business_process_modeling. Accessed 4 Sep 2021.

8 "Capital Efficiency Definition." Law Insider. https://www.lawinsider.com/dictionary/capital-efficiency. Accessed 10 Apr 2021.

9 "Computer Security Resource Center. NIST. https://csrc.nist.gov/glossary/term/commercial_off_the_shelf. Accessed 10 Sep 2021.

10 "Science is Never Settled." https://scienceisneversettled.com/consensus-is-not-science/. Accessed 9 Sep 2021.

11 "Learn About Quality." ASQ. https://asq.org/quality-resources/continuous-improvement. Accessed 23 Aug 2021.

12 "Creative destruction." MIT. https://economics.mit.edu/files/1785. Accessed 1 May 2021.

13 Kramer, S. W. & Jenkins, J. L. (2006). Understanding the Basics *of CPM Calculations: What is Scheduling Software Really Telling You? Paper presented at PMI® Global Congress 2006 – North America, Seattle, WA. Newtown Square, PA: Project Management Institute.*

14 "Culture." Merriam-Webster.com Dictionary, Merriam-Webster, https://www.merriam-webster.com/dictionary/culture. Accessed 30 Aug 2021.

15 "What Is Cybersecurity?" Cisco. https://www.cisco.com/c/en/us/products/security/what-is-cybersecurity.html. Accessed 29 Aug 2021.

16 "Three types of bias: Distortion of research results and how that can be prevented." PubMed.gov. https://pubmed.ncbi.nlm.nih.gov/25714762/. Accessed 1 Sep 2021.

17 "Data reduction." Wikipedia. https://en.wikipedia.org/wiki/Data_reduction. Accessed 18 Sep 2021.

18 "Deterministic." Dictionary.com. https://www.dictionary.com/browse/deterministic. Accessed 1 May 2021.

19 "Digitalization." Merriam-Webster.com Dictionary, Merriam-Webster, https://www.merriam-webster.com/dictionary/digitalization. Accessed 10 Jul 2021.

20 "What is economic cost?" Market Business News. https://marketbusinessnews.com/financial-glossary/economic-cost/. Accessed 26 Aug 2021.

21 "Marginal utility." Britannica. https://www.britannica.com/topic/marginal-utility. Accessed 27 Aug 2021.
22 "Utility." Investopedia. https://www.investopedia.com/terms/u/utility.asp. Accessed 22 Aug 2021.
23 "Ecosystem." Britannia. https://www.britannica.com/science/ecosystem. Accessed 31 Aug 2021.
24 "Empirical." Merriam-Webster.com Dictionary, Merriam-Webster, https://www.merriam-webster.com/dictionary/empirical. Accessed 18 Sep 2021.
25 "Endogenous." Merriam-Webster.com Dictionary, Merriam-Webster, https://www.merriam-webster.com/dictionary/endogenous. Accessed 9 Jul 2021.
26 "Enterprise Resource Planning (ERP)." Investopedia. https://www.investopedia.com/terms/e/erp.asp. Accessed 29 Aug 2021.
27 "Exogenous." Merriam-Webster.com Dictionary, Merriam-Webster, https://www.merriam-webster.com/dictionary/exogenous. Accessed 9 Jul 2021.
28 "Expected Value (EV)." Investopedia. Financial Analysis. https://www.investopedia.com/terms/e/expected-value.asp. Accessed 21 Sep 2021.
29 Shemwell, Scott M. (1997, September). The Economic Value of Timely Information and Knowledge, Key to Business Process Integration Across Boundaries in the Oil & Gas Extended Value Chain. Proceedings *of the Gulf Publishing 3rd International Conference and Exhibition on Exploration & Production Information Management. Houston.*
30 "Introducing FBLearner Flow: Facebook's AI backbone." Facebook Engineering. https://engineering.fb.com/2016/05/09/core-data/introducing-fblearner-flow-facebook-s-ai-backbone/. Accessed 11 Oct 2021.
31 "5G." Wikipedia. https://en.wikipedia.org/wiki/5G. Accessed 17 Sep 2021.
32 "What Is the Fourth Industrial Revolution?" The 360 blog. Salesforce. https://www.salesforce.com/blog/what-is-the-fourth-industrial-revolution-4ir/. Accessed 27 Sep 2021.
33 "Future-proof." Collinsdictionary.com. Collins. Accessed 23 Aug 2021. https://www.collinsdictionary.com/us/dictionary/english/future-proof.
34 "Generally Accepted definition." Law Insider. https://www.lawinsider.com/dictionary/generally-accepted. Acc*essed 7 Sep 2021.*
35 "Harmonization (standards)." Wikipedia. https://en.wikipedia.org/wiki/Harmonization_(standards). Accessed 24 Sep 2021.
36 "HSE." Wikipedia. https://en.wikipedia.org/wiki/HSE. Accessed 18 Sep 2021.
37 "What is Human Factors and Ergonomics?" Human Factors and Ergonomics Society. https://www.hfes.org/About-HFES/What-is-Human-Factors-and-Ergonomics. Accessed 1 Sep 2021.
38 "Innovation." Merriam-Webster.com Dictionary, Merriam-Webster, https://www.merriam-webster.com/dictionary/innovation. Accessed 23 Aug 2021.
39 "What is Intellectual Property?" World Intellectual Property Organization (WIPO). https://www.wipo.int/about-ip/en/. Accessed 24 Sep 2021.
40 "Key Performance Indicators (KPIs)." Investopedia. https://www.investopedia.com/terms/k/kpi.asp. Accessed 23 Aug 2021.
41 "Learning management system." Wikipedia. https://en.wikipedia.org/wiki/Learning_management_system. Accessed 30 Sep 2021.
42 "Legacy Application Or System." Gartner Glossary. https://www.gartner.com/en/information-technology/glossary/legacy-application-or-system. Accessed 29 Aug 2021.
43 "Marginal cost." Wikipedia. https://en.wikipedia.org/wiki/Marginal_cost. Accessed 24 Aug 2021.

44 "Why Facebook's premature talk about a metaverse will likely backfire." Fast Company. https://www.fastcompany.com/90681568/facebook-metaverse-zuckerberg. Accessed 19 Oct 2021.

45 "What is Mindfulness?" Mindful. https://www.mindful.org/what-is-mindfulness/. Accessed 22 Sep 2021.

46 "What Does Minimum Viable Product (MVP) Mean? techopedia. https://www.techopedia.com/definition/27809/minimum-viable-product-mvp. Accessed 10 Sep 2021.

47 "Mission Critical." Investopedia. https://www.investopedia.com/terms/m/mission-critical.asp. Accessed 4 Sep 2021.

48 "ATT&CK®" MITRE. https://attack.mitre.org/. Accessed 18 Sep 2021.

49 "Net Promoter Score." Wikipedia. https://en.wikipedia.org/wiki/Net_promoter_score. Accessed 24 Aug 2021.

50 "Business Process Maturity Model, v1.0." Object Management Group. https://www.omg.org/spec/BPMM/1.0/PDF. Accessed 25 Aug 2021.

51 "One throat to choke." TechTarget. https://searchconvergedinfrastructure.techtarget.com/definition/one-throat-to-choke#:~:text=One%20throat%20to%20choke%20is,%E2%80%9Cone%20throat%20to%20choke.%E2%80%9D. Accessed 25 Aug 2021.

52 "Open Source." Wikipedia. https://en.wikipedia.org/wiki/Open_source. Accessed 29 Aug 2021.

53 "Operating Cost." Investopedia. https://www.investopedia.com/terms/o/operating-cost.asp. Accessed 23 Aug 2021.

54 "Operational Excellence Platform." The Rapid Response Institute. https://therrinstitute.com/operational-excellence-platform/. Accessed 23 Aug 2021.

55 "Opportunity Cost." The Library of Economics and Liberty. https://www.econlib.org/library/Enc/OpportunityCost.html. Accessed 5 Apr 2021/.

56 Performance Management: Concepts & Definitions." Hr.berkeley.edu. UK Berkeley. https://hr.berkeley.edu/hr-network/central-guide-managing-hr/managing-hr/managing-successfully/performance-management/concepts. Accessed 23 Aug 2021.

57 "Project Management Body of Knowledge." Wikipedia. https://en.wikipedia.org/wiki/Project_Management_Body_of_Knowledge. Accessed 30 Sep 2021.

58 "What is portfolio management? Association of Project Management. https://www.apm.org.uk/resources/what-is-project-management/what-is-portfolio-management/. Accessed 17 Sep 2021.

59 "Predictive maintenance." Wikipedia. https://en.wikipedia.org/wiki/Predictive_maintenance. Accessed 17 Sep 2021.

60 "Proof of Concept: Definition & Best Practices." Project Manager. https://www.project-manager.com/blog/proof-of-concept-definition. Accessed 27 Sep 2021.

61 "What is a Reference Architecture?" Hewlett Packard Enterprise. https://www.hpe.com/us/en/what-is/reference-architecture.html. Accessed 21 Sep 2021.

62 "Reliability." Merriam-Webster.com Dictionary, Merriam-Webster, https://www.merriam-webster.com/dictionary/reliability. Accessed 9 Jul 2021.

63 "R & D." Collins English Dictionary. collinsdictionary.com. https://www.collinsdictionary.com/us/dictionary/english/r-and-d. Accessed 25 Aug 2021.

64 "A Guide to Calculating Return on Investment (ROI)." Investopedia. https://www.investopedia.com/articles/basics/10/guide-to-calculating-roi.asp. Accessed 25 Aug 2021.

65 "Right to operate definition." Law Insider. https://www.lawinsider.com/dictionary/right-to-operate#:~:text=Right%20to%20operate%20means%20that,The%20solid. Accessed 24 Sep 2021.

66 "Risk management." Wikipedia. https://en.wikipedia.org/wiki/Risk_management. Accessed 23 Aug 2021.

67 "Sarbanes-Oxley (SOX) Act of 2002." Investopedia. https://www.investopedia.com/terms/s/sarbanesoxleyact.asp. Accessed 29 Sep 2021.

68 "Selling, General & Administrative Expense (SG&A)." Investopedia. https://www.investopedia.com/terms/s/sga.asp. Accessed 17 Sep 2021.

69 "Single sign-on." Wikipedia. https://en.wikipedia.org/wiki/Single_sign-on. Accessed 25 Aug 2021.

70 "Situation awareness." APA Dictionary of Psychology. American Psychological Association. https://dictionary.apa.org/. Accessed 30 Sep 2021.

71 "What is Smart Technology." IGI Global. https://www.igi-global.com/dictionary/smart-interactive-game-based-system-for-preschools-in-tanzania/38186. Accessed 23 Aug 2021.

72 Tassey, G. (2000). "Standardization in technology-based markets." Research Policy 29(4–5): 587–602.

73 "STEM." Merriam-Webster.com Dictionary, Merriam-Webster, https://www.merriam-webster.com/dictionary/stem. Accessed 31 Aug 2021.

74 "Stochastic" Dictionary.com. https://www.dictionary.com/browse/stochastic. Accessed 24 Aug 2021.

75 "Subject Matter Expert (SME) Roles and Responsibilities." Mississippi Accountability System for Government Information and Collaboration (MAGIC). https://www.dfa.ms.gov/media/9207/subject-matter-expert-sme-roles-and-responsibilities.pdf. Accessed 19 Sep 2021.

76 "SCADA." Wikipedia. https://en.wikipedia.org/wiki/SCADA. Accessed 7 Sep 2021.

77 "Supply Chain Management (SCM)." Investopedia. https://www.investopedia.com/terms/s/scm.asp. Accessed 29 Aug 2021.

78 "Business Use Case." University of Houston Clear Lake. https://sceweb.uhcl.edu/helm/RationalUnifiedProcess/process/modguide/md_buc.htm. Accessed 15 Jul 2021.

79 "Validity." Merriam-Webster.com Dictionary, Merriam-Webster, https://www.merriam-webster.com/dictionary/validity. Accessed 9 Jul 2021.

80 "Rapid Response Management: Thriving in the New World Order." The Rapid Response Institute. https://therrinstitute.com/wp-content/uploads/2017/10/rapid_response_management_-_thriving_in_the_new_world_order_2.0_-_january_2009.pdf. Accessed 2 Jun 2021.

81 "Money Velocity." FRED Economic Data. https://fred.stlouisfed.org/categories/32242. Accessed 2 Jul 2021.

82 "Wisdom of Crowds. Investopedia. What is Wisdom of Crowds? https://www.investopedia.com/terms/w/wisdom-crowds.asp. Accessed 17 Sep 2021.

83 "Workflow engine." Wikipedia. https://en.wikipedia.org/wiki/Workflow_engine. Accessed 4 Sep 2021.

84 Rutherford, Donald. (2012). *Dictionary of Economics*. London: Routledge.

Appendix 2
Industry Organizations and Standards Bodies

The following key industry organizations are used in this report and by many practitioners. These definitions are provided for the convenience of the readers. This list is not all inclusive and local regulations and practices may dictate additional and/or other standards and practices. The intent is to document that Smart Manufacturing efforts may cross multiple standards and practices.

Note: As a general rule, the following descriptions are taken from the relevant website and are references as appropriate. As used throughout this report, direct quoted citations are in *italics*, per editorial convention.

Industry Organizations and Standards Bodies

Organization	Description
American Institute of Chemical Engineers (AIChE)	*The world's leading organization for chemical engineering professionals, with more than 60,000 members from more than 110 countries.*[1]
American National Standards Institute (ANSI)	*A private, nonprofit organization that administers and coordinates the U.S. voluntary standards and conformity assessment system. The Institute provides a framework for fair standards development and quality conformity assessment systems and continually works to safeguard their integrity.*[2]
American Society of Mechanical Engineers (ASME)	*Promotes the art, science, and practice of multidisciplinary engineering and allied sciences around the globe.*[3]
ASTM International	*Over 12,000 ASTM standards operate globally. Defined and set by us, they improve the lives of millions every day. Combined with our innovative business services, they enhance performance and help everyone have confidence in the things they buy and use.*[4]

(Continued)

Organization	Description
CESMII—The Smart Manufacturing Institute	*CESMII MISSION—Radically accelerates the development and adoption of advanced sensors, controls, platforms, and models, to enable Smart Manufacturing (SM) to become the driving sustainable engine that delivers real-time business improvements in U.S. manufacturing.*[5]
Committee of Sponsoring Organizations (COSO)	*Mission is to help organizations improve performance by developing thought leadership that enhances internal control, risk management, governance, and fraud deterrence.*[6]
International Electrotechnical Commission (IEC)	*The IEC is a global, not-for-profit membership organization, whose work underpins quality infrastructure and international trade in electrical and electronic goods. Our work facilitates technical innovation, affordable infrastructure development, efficient and sustainable energy access, smart urbanization and transportation systems, climate change mitigation, and increases the safety of people and the environment.*[7]
International Organization for Standardization (ISO)	*Develops and publishes International Standards.*[8]
International Society of Automation (ISA)	*A nonprofit professional association founded in 1945 to create a better world through automation.*[9]
Internet Engineering Task Force (IEEE)	*The world's largest technical professional organization dedicated to advancing technology for the benefit of humanity.*[10]
Internet Engineering Task Force (IETF)	*The mission of the IETF is to make the Internet work better by producing high-quality, relevant technical documents that influence the way people design, use, and manage the Internet.*[11]
ISACA (formerly the Information Systems Audit and Control Association)	*An international professional association focused on IT (information technology) governance.*[12]
MTConnect Institute	*A 501(c)(6) not-for-profit standards development organization for the MTConnect standard (ANSI/MTC1.4–2018). Its membership is made up of over 400 companies and research organizations in discrete manufacturing including automotive, aerospace, medical, and other industries as well as software developers, system integrators, and research organizations supporting those industries.*[13]
National Electrical Manufacturers Association (NEMA)	*An ANSI-accredited Standards Developing Organization made up of business leaders, electrical experts, engineers, scientists, and technicians.*[14]

(Continued)

Organization	Description
National Institute of Standards and Technology (NIST)	*Part of the U.S. Department of Commerce. NIST is one of the nation's oldest physical science laboratories.*[15]
Object Management Group (OMG)	*An open membership, not-for-profit computer industry standards consortium that produces and maintains computer industry specifications for interoperable, portable, and reusable enterprise applications in distributed, heterogeneous environment.*[16]
Project Management Institute (PMI)	*The world's leading professional association for a growing community of millions of project professionals and changemakers worldwide.*[17]
RAPID Manufacturing Institute	*A Public–Private Partnership Between U.S. DOE EERE, and AIChE.*[18]
Smart Manufacturing Leadership Coalition (SMLC)	*Focus on cultivating trusted and mutually beneficial collaborative activities for member organizations of all sizes by sharing successes and challenges related to the adoption of smart manufacturing technologies and 4th Industrial Revolution digital technologies.*[19]
The French Association of Internet Users (AFNeT)	*Network of experienced and recognized players has been at the service of industrial sectors to support and develop projects such as the European collaborative hub BoostAeroSpace or the international PLM standards for aeronautics and the automobile industry.*[20]
World Wide Web Consortium (WC3)	*The mission is to lead the World Wide Web to its full potential by developing protocols and guidelines that ensure the long-term growth of the Web.*[21]

NOTES

1 "About AIChE." AIChE. https://www.aiche.org/about. Accessed 15 Aug 2021.
2 "About ANSI." ANSI. https://ansi.org/about/introduction. Accessed 15 Aug 2021.
3 "Setting the Standard." The American Society of Mechanical Engineers. https://www.asme.org/. Accessed 1 Sep 2021.
4 "About Us." ASTM International. https://www.astm.org/ABOUT/overview.html. Accessed 1 Sep 2021.
5 "CESMII – The Smart Manufacturing Institute." CESMII. https://www.cesmii.org/. Accessed 15 Aug 2021.
6 "Committee of Sponsoring Organizations of the Treadway Commission." COSO. https://www.coso.org/Pages/aboutus.aspx. Accessed 27 Sep 2021.
7 "What-We-Do." International Electrotechnical. Commission. https://www.iec.ch/what-we-do. Accessed 16 Aug 2021.

8 "ISO." International Organization for Standardization. https://www.iso.org/home. html. Accessed 16 Aug 2021.

9 "Setting the Standard for Automation." About ISA. https://www.isa.org/about-isa. Accessed 16 Aug 2021.

10 "Mission & Vision." IEEE. https://www.ieee.org/about/vision-mission.html. Accessed 15 Aug 2021.

11 "Mission and Principles." Internet Engineering Task Force. https://www.ietf.org/ about/mission/. Accessed 1 Sep 2021.

12 "ISACA." Wikipedia. https://en.wikipedia.org/wiki/ISACA. Accessed 19 Sep 2021.

13 "MTConnect Institute." MTConnect. https://www.mtconnect.org/about. Accessed 24 Sep 2021.

14 "About the National Electrical Manufacturers Association." NEMA. https://www. nema.org/about. Accessed 16 Aug 2021.

15 "About NIST." NIST. https://www.nist.gov/about-nist. Accessed 16 Aug 2021.

16 "Business Process Maturity Model, v1.0." Object Management Group. https://www. omg.org/spec/BPMM/1.0/PDF. Accessed 25 Aug 2021.

17 "About Us." Project Management Institute. https://www.pmi.org/about. Accessed 19 Sep 2021.

18 "Roadmap Overview." AIChE. https://www.aiche.org/sites/default/files/docs/ pages/2008_17_rapid_roadmap_booklet.pdf. Accessed 16 Aug 2021.

19 "Accelerating the Digital Transformation of Manufacturers and their Supply Chains." Smart Manufacturing Leadership Consortium. https://smlconsortium. org/. Accessed 24 Sep 2021.

20 "Accelerator of the Digital Transformation of Industrial Sectors." The French Association of Internet Users. https://www.afnet.fr/. Accessed 16 Aug 2021.

21 "W3C Mission." W3C. https://www.w3.org/Consortium/mission. Accessed 1 Sep 2021.

Appendix 3
Relevant Standards

The following standards are used in this report and by practitioners. These definitions are provided for the convenience of the readers.

Note: As a general rule, the following descriptions are taken from the relevant website and are references as appropriate. As used throughout this report, direct quoted citations are in *italics*, per editorial convention.

Standards

Standard	Description
AFNeT	*Smart Manufacturing Standards Map.*[1]
CESMII's Integrated Roadmap	*Webinar explores ways in which smart manufacturing and process development can come together to improve the energy efficiency and competitiveness of the chemical, oil and gas, and bioproducts industries.*[2]
ISA95, Enterprise-Control System Integration	*A standard that will define the interface between control functions and other enterprise functions based upon the Purdue Reference Model for CIM.*[3]
ISO 31000:2018 Risk management—Guidelines	*Guidelines on managing risk faced by organizations. The application of these guidelines can be customized to any organization and its context.*[4]
ISO 9001:2015	*Specifies requirements for a quality management system when an organization.*[5]
ISO Management System Standards List	*Management Standards (MS) support governance and leadership functions, at all levels. They are designed to be widely applicable across economic sectors (or specific to some), various types and sizes of organizations, and diverse geographical, cultural, and social conditions. MS can be considered as overarching documents for the sound governance of an organization.*[6]

(Continued)

Standard	Description
ISO/IEC TR 63306-1:2020	*This document describes the framework and the vocabulary that are used for the development of entries in the Smart Manufacturing Standards Map Catalogue. These enable the mapping and linking of standards and standard projects related to various aspects of smart manufacturing.*[7]
ISO 31030:2021	*Travel risk management—Guidance for organizations.*[8] (might be useful for organizations with manufacturing facilities in high-risk regions).
ISO 45000 Family: Occupational Health and Safety	*Improving employee safety, reducing workplace risks, and creating better, safer working conditions.*[9]
NISTIR 8107: Current Standards Landscape for Smart Manufacturing Systems	*This report provides a review of the body of pertinent standards—a standards landscape— upon which future smart manufacturing systems will rely.* Readers should note that this report was published in 2016 and may be somewhat dated.[10]
Occupational Safety and Health Administration (OSHA)	*Compliance assistance resources tailored to specific industries, including those listed below. These resources include eTools and Safety and Health Topics pages. To find additional compliance assistance resources for your industry.*[11]
Smart Manufacturing and Standards: The NIST Role	*NIST Contribution—Measurement science and standards to drive innovation and reduce risks of adoption of Smart Manufacturing Technologies.* Readers should note that this report was published in 2016 and may be somewhat dated.[12]
Smart Manufacturing Reference Architecture	*Reference architecture model, i.e., a uniform conceptual and methodological structure, forms a basis for ensuring that the experts involved from the various disciplines master this complexity and speak a common language. It creates a common structure for the uniform description and specification of concrete system architectures.*[13]
The IEEE CS Standards Activities	Video overview of the IEEE Smart Manufacturing Standards.[14]

NOTES

1 "Smart Manufacturing Standards Map." AFNeT Standards Days 2021: 1 and 2 June 2021. https://www.afnet.fr/Content/2021–06–0102-ASD-Programme/2021–06-02-prez-PDF/ASD2021-11-JosephBriant-m0531-18H13.pdf. Accessed 17 Aug 2021.

2 "Smart Manufacturing for the Process Industries." AIChE. https://www.aiche.org/academy/webinars/smart-manufacturing-process-industries. Accessed 12 Jul 2021.

3 "ISA95, Enterprise-Control System Integration." International Society of Automation. https://www.isa.org/standards-and-publications/isa-standards/isa-standards-committees/isa95. Accessed 24 Sep 2021.

4 "ISO 31000:2018(en)." ISO. https://www.iso.org/obp/ui#iso:std:iso:31000:ed-2:v1:en. Accessed 27 Sep 2021.

5 "Quality management systems – Requirements. ISO. https://www.iso.org/standard/62085.html. Accessed 9 Sep 2021.

6 "Management System Standards." ISO. https://www.iso.org/management-system-standards-list.html. Accessed 9 Sep 2021.

7 "ISO/IEC TR 63306-1:2020(en)." ISO. https://www.iso.org/obp/ui/#iso:std:iso-iec:tr:63306:-1:ed-1:v1:en. Accessed 17 Aug 2021.

8 "ISO 31030:2021(en): Travel Risk Management – Guidance for Organizations." ISO. https://www.iso.org/obp/ui/#iso:std:iso:31030:ed-1:v1:en. Accessed 17 Sep 2021.

9 "ISO 45000 Family: Occupational Health and Safety. ISO. https://www.iso.org/iso-45001-occupational-health-and-safety.html. Accessed 24 Sep 2021.

10 "Current Standards Landscape for Smart Manufacturing Systems." NIST. https://nvlpubs.nist.gov/nistpubs/ir/2016/NIST.IR.8107.pdf. Accessed 17 Aug 2021.

11 "Industry-Specific Resources." Occupational Safety and Health Administration. United States Department of Labor. https://www.osha.gov/complianceassistance/industry. Accessed 24 Sep 2021.

12 "Smart Manufacturing and Standards: The NIST Role." NIST. https://sites.nationalacademies.org/cs/groups/pgasite/documents/webpage/pga_175019.pdf. Accessed 17 Aug 2021.

13 "Reference Architecture Model Industrie 4.0." Standardization Council. https://www.sci40.com/english/rami4-0/. Accessed 17 Aug 2021.

14 "The IEEE CS Standards Activities Board." IEEE. https://www.youtube.com/watch?v=BPV7XHQqNVg. Accessed 17 Aug 2021.

Appendix 4
Research, Advisory, and Certification Services

The following service providers are used in this report and by practitioners. These definitions are provided for the convenience of the readers.

Note: As a general rule, the following descriptions are taken from the relevant website and are references as appropriate. As used throughout this report, direct quoted citations are in *italics*, per editorial convention.

Industry Services

Organization	Description
ABS Group	*Inspecting and verifying technology and equipment designs; delivering safety, risk, and compliance services; optimizing asset performance; providing advanced engineering support and certifying management systems for the marine, offshore, oil, gas, chemical, government and power sectors.*[1]
American Society of Safety Professionals (ASSP)	*For more than 100 years, we have supported occupational safety and health (OSH) professionals in their efforts to prevent workplace injuries, illnesses, and fatalities. We provide education, advocacy, standards development, and a professional community to our members in order to advance their careers and the OSH profession as a whole.*[2]
Bureau Veritas	*A world leader in testing, inspection, and certification services (TIC).*[3]
Collaborative for Academic, Social, and Emotional Learning (CASEL)	*Helping make evidence-based social and emotional learning an integral part of education from preschool through high school.*[4]
Committee of Sponsoring Organizations (COSO)	*Mission is to help organizations improve performance by developing thought leadership that enhances internal control, risk management, governance, and fraud deterrence.*[5]

(Continued)

Organization	Description
DNV	*Independent expert in assurance and risk management. Driven by our purpose, to safeguard life, property, and the environment, we empower our customers and their stakeholders with facts and reliable insights so that critical decisions can be made with confidence.[6]*
Gartner, Inc.	*World's leading research and advisory company and a member of the S&P 500. We equip business leaders with indispensable insights, advice, and tools to achieve their mission-critical priorities today and build the successful organizations of tomorrow.[7]*
IGI Global	*A leading international academic publisher committed to facilitating the discovery of pioneering research that enhances and expands the body of knowledge available to the research community.[8]*
International Data Corporation (IDC)	*Offers global, regional, and local expertise on technology and industry opportunities and trends in over 110 countries. IDC's analysis and insight helps IT professionals, business executives, and the investment community to make fact-based technology decisions and to achieve their key business objectives. Founded in 1964, IDC is a wholly owned subsidiary of International Data Group (IDG), the world's leading tech media, data, and marketing services company.[9]*
Lloyd's Register Group	*One of the world's leading providers of professional services for engineering and technology—improving safety and increasing the performance of critical infrastructures worldwide.[10]*
SGS	*Recognized as the global benchmark for quality and integrity. Our 93,000 employees operate a network of 2,600 offices and laboratories.[11]*
Social and Emotional Learning (SEL)	*An umbrella term to represent a wide array of nonacademic skills that individuals need in order to set goals, manage behavior, build relationships, and process and remember information.[12]*
Southwest Research Institute (SWRI).	*Leader among independent, nonprofit research and development organizations. Our staff of approximately 3,000 scientists, engineers, analysts, and support staff members continues to accomplish outstanding fundamental and applied engineering and research for clients from diverse segments of government and industry.[13]*

NOTES

1 "Our Solutions." ABS Group. https://www.abs-group.com/What-We-Do/. Accessed 9 Sep 2021.
2 "Welcome to ASSP." ASSP. https://www.assp.org/about/. Accessed. 9 Sep 2021.
3 "Welcome to Bureau Veritas." Bureau Veritas. https://group.bureauveritas.com/news-room/acquisition-aet-france-specializing-laboratory-testing-product-development-and. Accessed 9 Sep 2021.
4 "Advancing Social and Emotional Learning." CASEL. https://casel.org/. Accessed 1 Oct 2021.
5 "About Us." Committee of Sponsoring Organizations' (COSO). https://www.coso.org/Pages/aboutus.aspx. Accessed 22 Sep 2021.
6 "About Us." DNV. https://www.dnv.com/about/index.html. Accessed 9 Sep 2021.
7 "Gartner." https://investor.gartner.com/home/default.aspx#. Accessed 9 Sep 2021.
8 "About IGI Global." https://www.igi-global.com/about/. Accessed 9 Sep 2021.
9 "Premier Global Provider of Market Intelligence, Advisory Services, and Events." International Data Corporation (IDC). https://www.idc.com/about. Accessed 19 Sep 2021.
10 "About us." Lloyd's Register. https://www.lr.org/en-us/who-we-are/. Accessed 9 Sep 2021.
11 "About SGS." SGS. https://www.sgs.com/en/our-company/about-sgs. Accessed 9 Sep 2021.
12 "About Explore SEL." Harvard University. Explore SEL. http://exploresel.gse.harvard.edu/about/. Accessed 1 Oct 2021.
13 "President's Message." Southwest Research Institute. https://www.swri.org/who-we-are/presidents-message. Accessed 9 Sep 2021.

Appendix 5
Commercial Tools and Services

There is a large number of providers of financial, project, and ROI calculators as well as empirical project databases, and it is not possible to include them all in this book. Therefore, the following are representative of those providing these services. As before, the publishers and authors are not recommending any given firm or product—caveat emptor.

The following tools are offered by various commercial services which may charge for the tool, data, and/or professional services and assessment. These resources are provided for the convenience of the readers.

Note: As a general rule, the following descriptions are taken from the relevant website and are references as appropriate. As used throughout this report, direct quoted citations are in *italics*, per editorial convention.

Resource	Description
Self-Assessment of Your Organization's Culture of Safety Maturity	*Where does your company's Culture of Safety (COS) rank against industry peers?[1]*
Smart Manufacturing Maturity Checklist	*This assessment includes the Industrial Network, equipment, processes, and people.[2]*
Worldwide Digital Transformation Spending Guide	*Comprehensive database delivered via IDC's Customer Insights query tool allows the user to easily extract meaningful information about the digital transformation market by viewing data trends and relationships and making data comparisons.[3]*

NOTES

1 "Self-Assessment of Your Organization's Culture of Safety Maturity." The Rapid Response Institute. https://therrinstitute.com/oe-solution-set-2/maturity-models/the-cos-maturity-model/cos-maturity-self-assessment/cos-maturity-conducting-your-self-assessment/. Accessed 20 Sep 2021.
2 "Smart Manufacturing Maturity Checklist." Polytron. https://polytron.com/resources-insights/smart-manufacturing-maturity-checklist/. Accessed 20 Sep 2021.
3 "Worldwide Digital Transformation Spending Guide." IDC. https://www.idc.com/tracker/showproductinfo.jsp?containerId=IDC_P32575. Accessed 19 Sep 2021.

References

Boebert, Earl and Blossom, James M. (2017). *Deepwater Horizon: A System Analysis of the Macondo Disaster*. Cambridge, MA: Harvard University Press.

Holland, Winford "Dutch" E. and Shemwell, Scott M. (2014). *Implementing a Culture of Safety: A Roadmap to Performance-Based Compliance*. New York: Xlibris.

Kramer, S. W. & Jenkins, J. L. (2006). *Understanding the Basics of CPM Calculations: What is Scheduling Software Really Telling You?* Paper presented at PMI® Global Congress 2006—North America, Seattle, WA. Newtown Square, PA: Project Management Institute.

Niven, Chris and Prouty, Kevin. (2017). *IDC TechBrief: Operations Management Systems*. IDC.

Quazi, Hebab A. (1992) *Cogeneration*, Published in Encyclopedia of Physical Science and Technology, Vol 3. Academic Press.

Quazi, Hebab A. (2020). *Commercializing Nanotechnology – A Roadmap to Taking Nanoproducts from Laboratory to Market*. CRC Press.

Rutherford, Donald. (2012). *Dictionary of Economics*. London: Routledge.

Shemwell, Scott M. (1993). Management Theory—Evolution Not Revolution. *Proceedings of the 11th Annual Conference of the Association of Management*, 11 (2), pp. 74–78.

Shemwell, Scott M. (1997, September). The economic value of timely information and knowledge, key to business process integration across boundaries in the Oil & Gas extended value chain. *Proceedings of the Gulf Publishing 3rd International Conference and Exhibition on Exploration & Production Information Management*. Houston.

Shemwell, Scott M. (2004, November). Governing Value. Governing Energy. *Executive Briefing – Business Value from Technology*. Vol 3 No. 11.

Shemwell, Scott M. (2011). *Essays on Business and Information II: Maximizing Business Performance*. New York: Xlibris. p. 13.

Shemwell, Scott M. (2012, March). *Integrity Management: Issues & Trends Facing the 21st Century Energy Industry*. The Rapid Response Institute.

Venkataramani, Swetha. (2021). *6 Key Actions for a Successful Smart Manufacturing Strategy, Smarter With Gartner*. August 23, 2021. https://www.gartner.com/smarterwithgartner/6-key-actions-for-a-successful-smart-manufacturing-strategy/.

Weiss, Stephen E. (1993, May). Analysis of Complex Negotiations in International Business: The RBC Perspective*. *Organization Science*. Vol. 4, No. 2.

Index

Note: **Bold** page numbers refer to tables; *italic* page numbers refer to figures.

Printed in the United States
by Baker & Taylor Publisher Services